# WHERE THE DESERT MEETS THE MOUNTAINS

## A Descriptive Geography of Western Imperial Valley and Eastern San Diego County

A Historical Perspective from the 1980s

Reynaldo Ayala, PhD

Marta Stiefel Ayala, PhD

**WHERE THE DESERT MEETS THE MOUNTAINS**



ISBN: 978-1-716-51588-0
Imprint:Lulu.com

# FOREWORD TO THE 2020 EDITION

This foreword is an introduction to the 2020 version of Where the Desert Meets the Mountains. The original document was completed in 1985 and was intended to be the 6th Occasional Paper on a series of publications planned by the Imperial Valley College Museum (IVCM). IVCM resides within the Imperial Valley College, a regional community college located in Imperial Valley, California.

This document was the result of an archeology field study class offered by Professor Jay von Werlhof during the late 1970s and early 1980s. Marta Ayala was a student in his class, and Reynaldo Ayala was her companion and assistant during these field trips. The final course assignment for this class was a paper that became the basis for this study. Professor von Werlhof encouraged Marta and Reynaldo to continue their research and complete a more comprehensive document to be published as part of the IVC Museum series.

The original document was approved for publication by the museum, yet it was not fully understood why this, and other documents, were not formally published as intended.

After many years, the original document was found inside a metal box in our residential garage. There were no available computers in 1985, thus the original document was typewritten on an electrical typewriter and the corresponding photos applied with paste. The map included in the original document was produced by hand, and typographical errors were edited using Liquid Paper Correction Fluid.

In preparing this new document, optical character recognition (OCR) was used via scanning to convert the original text into an editable word document. Fortunately, negatives of the original photographs were also discovered and subsequently scanned and edited using Photoshop. Only the appendices, bibliography, glossary, tables, and map were directly copied into this document.

This new document has been wonderfully edited for grammatical errors, typos, and inconsistencies by Cat Shehan Ayala. The original text, research, documentation, statistics, and photography were kept exactly as written and photographed in 1985.

The resulting document presented here is a truly historic one!

Reynaldo Ayala

# ACKNOWLEDGEMENTS

We dedicate this book to Irene Westling, con cariño y admiración.

We wish to thank Ruth Hutchison, SDSU-IVC, for her expert typing and the many hours spent on proofreading the material;  Steve Settimi for his help with the research and writing; Jaime Servin, IVC Museum, for his excellent work with the maps and graphics; Harry Casey, IVC Museum for his darkroom expertise, and Enrique Monroy, for supervising the printing of the entire work.

Special thanks to Jay von Werlhof, IVC instructor and chief curator of the IVC Museum, for encouraging us to finish this book.

The research for this monograph was made possible thanks to a San Diego State University Grant-in-Aid, 1977.

# TABLE OF CONTENTS

# TABLE OF PHOTOS

# PURPOSE

The regional geography of the Colorado Desert and the Peninsular Ranges Province has been thoroughly described by various authors. The heavily cultivated lands of the Imperial and Coachella Valleys have been under study since the end of the 19th century. But only recently, the extensive lands located between the old beach line and the eastern watershed of the Peninsular Ranges have come under scrutiny as public lands have come under pressure for utilization by developers and various interest groups. Before that happens, there is an urgent need for historical and anthropological study and research of this area, as it now appears that this region was extremely important for early man as he migrated from the Peninsular Ranges (Jacumba, In-Ko-Pah, Laguna, Volcan, and Santa Rosa Mountains) down to the desert floor using an infinite number of canyons and washes for pathways.

The nature of the surrounding environment is of extreme importance in the understanding of human occupancy patterns since "it allows for an interdisciplinary approach to the archaeology of the area. Many in-roads have been made into the use of such disciplines as geology, hydrology, geomorphology, palynology, soil science, paleontology, etc. for the purpose of eliciting past environments. This is especially important in light of problems encountered with the lack of absolute dating techniques applicable for early culture horizons" (USDl-BLM, Yuha Basin…, 1980, p. 38).

We can assume that several environmental elements should be considered as keys to the understanding of human occupancy in the study area. Some of these elements are water sources, lithic resources, distribution and availability of flora and fauna, and habitable lands. Therefore, the purpose of this work is to present a general understanding of the physical characteristics of this California landscape to serve as basic background for students, researchers, and users of this fragile land.

Even the earliest inhabitants of the desert, passive by today's approaches, brought change into the land. They flood-irrigated the land, mined and quarried for exotic and useful stones, and created a maze of foot trails for commerce and travel. The nature of the desert environment contributed to the preservation of these archaeological and historical sites. "To date over 12,000 archaeological sites have been recorded in the California deserts and no doubt more than thousands of California desert archaeological sites have been destroyed and thousands more damaged by human activity and natural events" (USDI-ELM, Draft…, 1980, p. 51).

Eleven different types of archaeological sites have been found in the study area. They are: 1. burial or cremation, 2. cairn or monument, 3. intaglios, 4. isolated find, 5. lithic scatter, 6. pottery scatter, 7. roasting pit or hearth, 8. sleeping circles, 9. temporary camp, 10. trails, and 11. village (USDl-BLM, Yuha Basin…, 1980. p. 40).

The danger of destruction of anthropological evidence by other land uses – recreation, off-road vehicles, and agriculture – requires that intensive fieldwork, analysis, and documentation take place before it is too late.

{ x }

# INTRODUCTION

Southern California is a region of great geologic and topographic diversity. The land in this region may be divided into four natural landform areas or provinces, mainly based on distinctive physiographic characteristics. This monograph will concern itself only with a portion of two of these natural regions: the Colorado Desert Province and the Peninsular Ranges Province.

The Colorado Desert Province (4000 square miles and part of the larger Sonoran Desert) comprises a series of sediment-filled, low-lying basins located between the Peninsular Ranges Province to the west, and the Little San Bernardino, Orocopia, and Chocolate Mountains to the east (Lentis, 1977, p. 58).

The Peninsular Ranges Province is a series of fault-block mountains with granitic cores and steep eastern faces extending into the Colorado Desert and forming the southwestern portion of the California Coastal Range. It extends southeastward to the Baja California Peninsular Range - the same physiographic unit divided only by the geopolitical boundary of the international border (Lentis, 1977, pp. 88-89).

This study is an attempt to describe the physiographic characteristics of an area within the two named provinces; limited to the east by the western shores of the Salton Sea and the irrigated lands of the Imperial Valley; to the west by the eastern watershed of the Peninsular Ranges (In-Ko-Pah, Jacumba, Laguna, Tierra Blanca and Vallecito Mountains); to the north by the southern watershed of the Santa Rosa Mountains, south of the Riverside County line; and to the south by the California-Baja California (Mexico) international boundary line. See map 1.

The geographical characteristics to be considered for the area under study are:
1. Geomorphology, 2. Climate, 3. Soils, 4. Biogeography, and 5. Human presence. Special emphasis shall be given to the description of those physical phenomena which help relate the story of early man in the area.

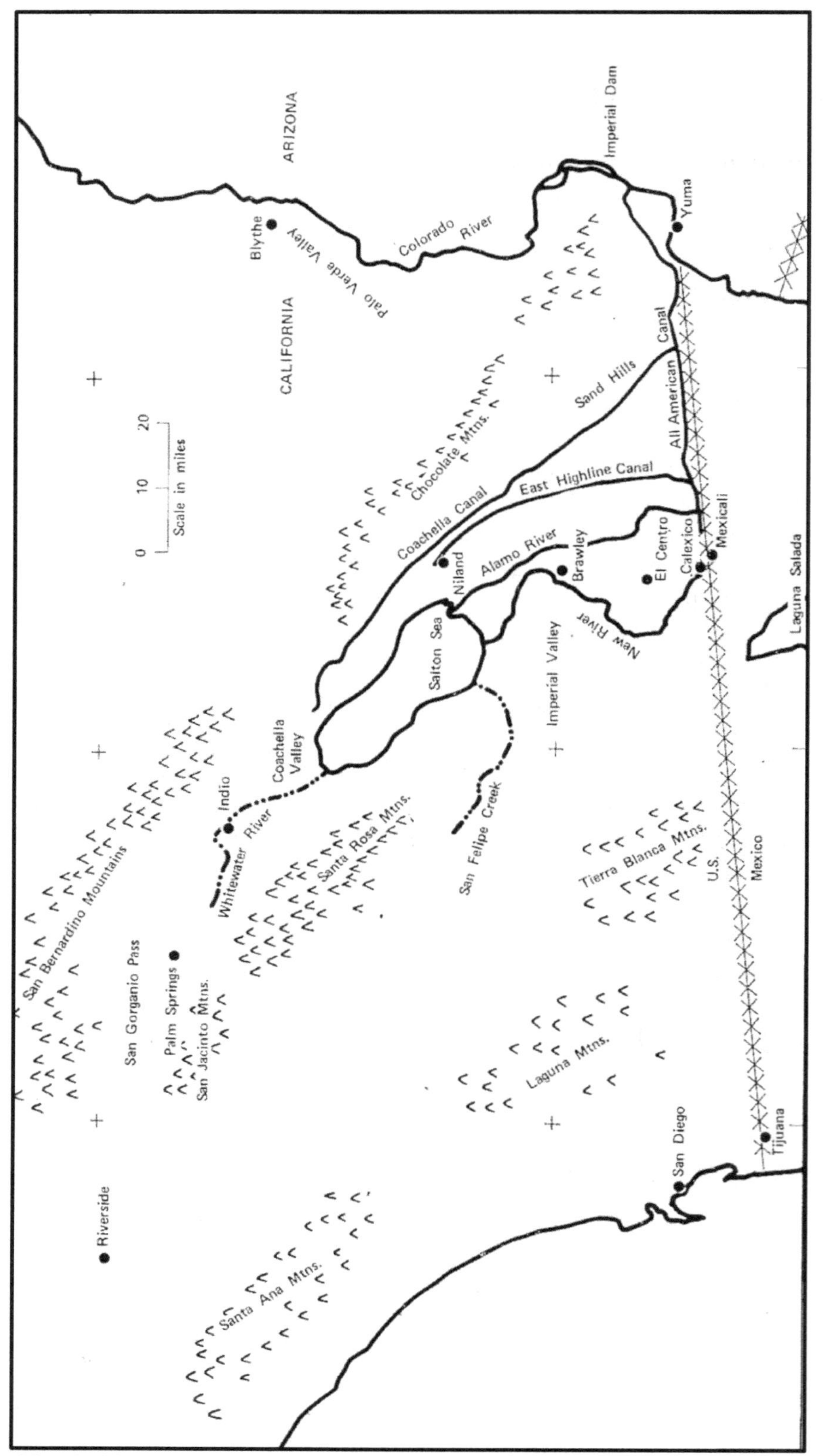


**Map 1 - Map of study area**

**CHAPTER I**

## The Colorado Desert Province

The Colorado Desert Province to a large extent coincides with the landform feature known as the Salton Basin. This trough is an elongated structural basin with a vast thickness of down-faulted sediments. It has its apex to the northwest, not far from where the San Bernardino and San Jacinto Mountains merge to form the San Gorgonio Pass. It extends toward the southeast into Mexico but becomes separated from the larger depression of the Gulf of California by the delta of the Colorado River. To the west, the basin is delimited by four abrupt peninsular projections which protrude into the province but are themselves part of the Peninsular Ranges. Its eastern limit is demarcated by the Little San Bernardino, Orocopia, and Chocolate Mountains (Lantis, 1977, pp. 58-59).

The basin is an area of internal drainage and receives run-off from approximately 9000 square miles of barren and mountainous terrain, as well as drainage from agricultural irrigated districts. The highest parts of the basin, on the western edge of the depression, have a marked fall toward the north-west-southeast axis of the trough. The washes, which have their origins in the Santa Rosa, Coyote, Vallecito, and Jacumba Mountains, flow east and northeast into the basin. The drainage of the region reflects the characteristic drainage patterns of stream courses in arid lands: wide, shallow, sandy-bottomed channels often having poorly defined banks. These intermittent water courses are dry most of the time and carry water only for brief periods after long and heavy rains in the higher elevations or adjacent desert areas. None of this water reaches the large streams; the surface sands within a short distance absorb most of it. The basin is also laced with numerous creeks, gullies, and arroyos, including the deep gorges of the New and Alamo Rivers, which serve as drainage for surplus irrigation water and run-off.

The central and lowest portion of this immense graven, which at its deepest point is 232 feet below sea level, is occupied by the Salton Sea. Its highest elevations occur at the base of the west bordering mountains approximately 500 feet above sea level, and in the isolated prominences on the desert surface, which vary from 500 to 2000 feet above sea level. The northern portion of the basin, which slopes southward into the Salton Sea, is known as the Coachella Valley, an alluvium-filled valley with considerable irrigation and agriculture. The southern portion of the trough, which slopes northwest toward the Salton Sea, is called the Imperial Valley. It has large, irrigated areas devoted to agriculture.

Topographic Features
The basin is a generally flat, sedimentary landfill. The old beach line, occasional sand dunes, and partially buried mountain crops are the only conspicuous features on the desert floor. This work will discuss only those topographic features located within the area of interest; namely, the western portion of the Colorado Desert.

*The Ancient Beach Line*

One of the most notable topographic features of the Salton Basin is the nearly continuous old beach line which stands approximately 35 to 50 feet above sea level, marking the shores of ancient Lake Cahuilla. Presently, it encircles part of the Coachella Valley south of the city of Indio and the Imperial Valley, including the Salton Sea. It extends south into Baja California, Mexico. The sea level contour of the Salton Basin generally runs somewhat parallel to the old shoreline, approximately one-half mile east of it. However, there are portions of the old beach line which are found at angles to the sea level contour. This is evident in places such as the eastern end of the Lower Borrego Valley and south of Superstition Mountain and indicates the occurrence of faulting. See photos 1 and 2.

Photo 1 - Ancient Shoreline near Split Mountain. Fish Creek Mountains in the background.

The old beach line consists mainly of a sand ridge a few feet high, although its characteristics vary somewhat depending on the rocks upon which it was formed. Within the sand ridge there is an abundance of small, well-preserved freshwater shells. Near Travertine Palms, the shore of the ancient lake was once a granitoid mound; therefore, there are no beach deposits as such. Instead, the ancient shoreline is marked by incrustations of the massive lime deposit called travertine, which in places is several feet thick. This deposit, formed from the evaporation of lake water, appears as a white line visible from some distance away (Miller, 1957, p. 249). See photo 3.

**Photo 2 - Another view of the Ancient Shoreline, next to the rail line from U.S. Gypsum Co. Mine.**

All these elements point to the former existence of Lake Cahuilla, a large body of water, which appears to be one of the most recent features of this new, partially dry basin. The presence of non-marine shells suggests that originally, the water was fresh or nearly so. Lake Cahuilla, therefore, must have been formed in part by the Colorado River, which apparently discharged into the basin rather than into the Gulf of California. Overflows from the lake most likely drained into the gulf (Hinds, 1952, p. 103).

**Photo 3 - Ancient Shoreline in a spur of the Santa Rosa Mountains, close to Travertine Point.**

*Sand Dunes*
Southwest of the Salton Sea, between McCain Springs and Kane Springs, there is an area where small, crescent-shaped dunes called barchans are well developed. This type of sand dune evolves where a constant wind blows in the same direction and the supply of sand is somewhat limited. The barchans are asymmetrical when viewed in a cross section. Their gentle slope is oriented always in the direction of wind origin. Their steep side, as well as the horns of the crescent, are oriented in the opposite direction. The sand forming the barchans comes from an old beach nearby; practically all the sand has been consumed while forming the sand hills. Some of the barchans measure approximately 1000 feet from one point of the horn to the other, while others are no more than 300 feet long (Hinds, 1952, p. 100).

Another type of dune, such as those found in the San Felipe Valley, are the so-called "spring-formed dunes" which develop in sandy areas where seeping water permits the growth of vegetation; and thus the accumulation of sand. Sometimes, the sand piles so high that the water cannot penetrate through it and the spring becomes totally concealed. The core of the sand hill is a dark mass formed by decayed vegetation and soil. Kane Springs is an example of this type of formation. It occupies a mound approximately 30 feet high and several feet in diameter from which the water seeps eastward.

Another type of sand dune having a serpentine pattern occurs around Superstition Mountain. These are elongated and run parallel to the long axis of the mountain. They are found on both sides of the mountain and shift back and forth over its crest as the wind changes direction. Often, they blockade canyons forming temporary basins behind them.

In other parts of the basin, great numbers of irregular sand drifts have formed around clumps of vegetation, rock piles, and other natural obstructions.

*Superstition Mountain and Superstition Hills*
Superstition Mountain is a relatively low range; its highest point reaches approximately 760 feet above sea level. It is over 4 miles long and approximately 1 mile wide. It runs southeastwardly into the desert from a spur of the Fish Creek Mountains to which it was probably connected in ancient times. It is now separated by a gap approximately 4 miles wide through which the Carrizo Creek flows during the rainy season. It is composed of granite with non-fossiliferous sandstone on each side of the igneous rocks. In many places the sand has drifted entirely over the mountain and into the plain on the eastern side. The southwestern running slope, approximately 11.40º, has the most gradual gradient; the north facing slopes have an average gradient of 48% or a 26º slope. In the book, California Place Names, the following paragraph describes the mountain eloquently:

> While the hills appear to be composed entirely of sand, there is a rocky mass below, and over this the sand plays, constantly, shifting to and fro in the desert winds, and because of this instability the indians of the region speak ill of it, hence its name (Gudde, 1962, p. 309).

Superstition Hills are located north of Superstition Mountain and southwest of Highway 86. They are mudstone and sandstone hills littered with concretions, devoid of springs, and support little vegetation. Between Highway 78 and the hills are depressed plains, with mesquite-covered sand dunes, cut by deep washes which join to form the San Felipe Creek. This, in turn, drains into the Salton Sea. Located north of Superstition Hills is Harpers Well, an important stop for early travelers. Located southwest from Harpers Well, where the San Felipe and Carrizo Creeks merge, is the San Sebastian Marsh, an uncommon surfacing of fresh water in this dry land. It provides a habitat for many wildlife species including the desert pupfish. This was also the site in the 1900s for the would-be boomtown of San Felipe which never materialized (Parker, 1963, p. 46). See photo 4.

*Yuha Basin*
The Yuha Basin is located on the southwestern portion of the Salton Trough. This area appears mainly as a "nearly flat-lying desiccated lakebed plain" (USDI, Yuha Basin... 1980, p. 120). Four miles from the Jacumba Mountains, there is a slight rise of approximately 50 to 90 feet in the elevation of the plain. This rise, which marks the western extreme of the basin, gives the necessary impetus and starting point for the Yuha Wash, extending ten miles to the east and reaching the West Mesa. Yuha Wells are located on the north. Mount Signal, Sunrise Butte, and Yuha Buttes are the only topographic relief of significance within the basin. The Imperial Formation crops up in the Yuha Buttes and is overlain by the Palm Springs Formation composed of non-marine sandstone, hales, and clays. These two formations have created unstable slopes which have eroded into mini badland type topography (USDI, Yuha Basin... 1980, p. 120). See photo 5.

**Photo 4 – San Felipe Creek at the Junction of Highways S2 and 78.**

**Photo 5 - Badlands at Yuha Basin on Highway 98.**

*West Mesa*

Another desert plain, the Imperial West Mesa, lies on the west side of the old lakebed between the Fish Creek and Coyote Mountains to the west, and the Imperial Valley to the east. This nearly level plain includes the lower portions of the watershed of the Coyote Wash, Carrizo Creek, and San Felipe Creek. On this mesa are some remnants of old dissected fans which have slopes up to 3 percent. The fan sediments are the parent materials of the Antho, Laveen, Niland, and Superstition soils. Large areas of the West Mesa are sandy sediments deposited by intermittent streams from the Peninsular Ranges. Carsitas and Vint soils formed in the recently deposited sands, and Rositas and Superstition soils formed on older, more stable sandy deposits. Playas and areas of moderately fine or fine textured basin deposits are the source for the Glenbar, Holtville, and Imperial soils. At the edge of the playas and basins, and in the areas of deep aeolian dust deposits, are the silty materials in which the Indio soils formed.

The Naval Parachute Recovery Test Area lies 2.6 miles north of Plaster City, within the West Mesa, and is used for the purpose of target practice on a full-time basis. It is closed to the public.

The Carrizo Impact Area, to the west of the mesa, is a 27,000 acre no man's land which has been closed to the public most of the time since 1942. The Army leased the land from the State to use as a bombing range in 1942. Later, the Army leased the area to the Navy, which also used it for target practice. In 1959 the Navy sent special teams to clean the area of live ammunition and the area was opened to the public. Soon afterwards, civilians began to find live bombs, so the area was permanently closed to the public (Lindsay, 1978, p. 143).

*Lower Borrego Valley: Old Kane Spring Road*

The lower Borrego Valley, surrounded by the Yaqui and Pinyon Ridges, San Ysidro Mountains, Coyote Mountain, and Borrego Badlands, is the less confined southern extension of the Upper Borrego Valley. Following the Old Kane Spring Road (Pole Line Road) from the Narrows on Highway 78 of the Yaqui Ridge to the southeast, one runs into a bajada draining the northeast slopes of the Vallecito Mountains. After traversing many washes, the road reaches the base of the hills and continues across washes and wind-erased trails reaching the Gypsum Mine Road running between Ocotillo Wells and Split Mountain. The road continues through the abandoned town of Little Borrego, and eventually leads to one of the famous stands of elephant trees. The road reaches a great rift which forms the entrance to Split Mountain, the cliffs of which form the ridges of the Fish Creek Wash. Ore from the nearby United States Gypsum Mine is hauled by a narrow gauge railroad along the eastern base of the Fish Creek Mountains across the Carrizo Desert to Plaster City. On the hills bordering Fish Creek Wash to the right is the strontium mine worked for the manufacture of flares and fireworks (Parker, 1963, p. 52). See photo 6.

**Photo 6 - Rail line which connects gypsum mine (U.S. Gypsum) to Plaster City.**

The walls of Fish Creek became conglomerates and gave rise to a height of 200 feet. Until recently, it was a sluggish creek and its sandstone floor was covered with huge potholes full of water, some of them with fish. After the floods of 1916, the sandstone was covered with a thick layer of sand (Parker, 1963, p. 52).

**Photo 7 - Travertine Point.**

Travertine Subregion: Black Ravines
This subregion is located west of Highway 86 between the San Felipe Creek and the Tule Wash approximately one and a half miles west of the sand dunes. It encompasses an area of approximately 20 square miles. The Black Ravines have a general west-east trend and have a maximum average slope of nearly 15 degrees. It is an area of badlands, with scattered sandstone formations, concretions, fossils, and springs. Travertine Point is a mound of tufa-covered boulders; this calcareous tufa or travertine is a limestone deposit formed by algae when this mound was partially covered by Lake Cahuilla. Sometimes it is up to 18 inches thick. The old beach line is very evident here, as well as on the rocky projections of the Santa Rosa spur. Tule Springs is a year-round source of brackish water. See photo 7.

Just north of Travertine Point is Travertine Wash and Travertine Palms; the wash is choked with palo verdes and desert willows. Travertine Palms is an alkaline spring oasis with fan palms and date palms (Parker, 1963, p. 42). Located within the same area are the San Felipe Hills, formed by sandstone outcrops on a sedimentary bed. Many roads crisscross the area and there are remnants of "early wild cat oil drillings" (Parker, 1963, p. 44). Atop a huge sand dune are the Gas Domes, where gas bubbles and mud have formed miniature volcano-like mounds. Little vegetation grows here; only sparse clumps of creosote bush and mesquite.

Evolution of the Salton Basin
It is generally believed, as previously mentioned, that the Salton Trough is a down-faulted structural basin where diastrophism and subsequent alluvial fill have been the major modifiers of today's landscape.

The diastrophism took place in the late Cenozoic time (mainly Quaternary), as it is evident that the strata of this period was involved in the activity (Miller, 1957, p. 246). In some areas, the province rests upon metamorphic rocks of Mesozoic age; in other areas, vertical movement combined with lateral displacement has caused the stratigraphic section of the older rocks to be overturned.

Two major right-lateral fault zones, the San Jacinto and the Elsinore, transverse the area under study in a northwest-southeast direction.

The San Jacinto fault system, related to the San Andreas Fault system, is approximately 10 kilometers wide when it enters the trough, trends approximately 10 degrees northwest, and splits into three somewhat parallel strands west of the Imperial Valley. The northernmost strand extends through Clark Valley and the tip of the Santa Rosa Mountains. The middle strand extends into the Borrego Valley. The southernmost strand, the Coyote Creek fault, extends along the margins of the Borrego Mountains and Ocotillo Badlands to a point near the easternmost portion of Fish Creek Mountains (USDI, The Borrego Mountains..., 1972, p. 10). The Superstition Mountain and Superstition Hills' faults are undoubtedly part of this system (Biehler et al., 1964, p. 39). The San Jacinto fault system forms one of the major boundary faults along the western side of the Imperial Valley.

The Elsinore fault system is the southernmost, right-lateral fault, related to the San Andreas Fault system in the United States. It trends approximately 45 degrees northwest. It is a distinct, yet superficially discontinuous zone of fractures, extending southward from the Peninsular Ranges into the Salton Trough near Tierra Blanca Mountains (USDI Geological Survey, Borrego Earthquake..., 1972, p. 11). The vertical component on the fault can reach as high as 3000 feet, northwest of the Imperial Valley (Jahns, 1954, p. 45). Numerous subsidiary blocks and basins are aligned along these major strike slip faults.

On the southwestern side of the basin, there is a set of complex fault block splinters associated with the major systems which have created a series of rugged mountain spurs, such as the Fish Creek and Vallecito Mountains, projecting into the alluvial floor and creating deep valleys such as those of the San Felipe and Carrizo Creeks. In other places, the floor of the Salton Basin merges imperceptibly with the floor of the valleys of the Peninsular Ranges (Durrenberger, 1968, p. 19). See photo 8.

Several faults have been recognized near Warner Valley, which extend southward in this part of the region. Several valleys such as Borrego, San Felipe, Vallecito, Collins, and others were apparently formed by these faults. Each of these valleys has as its northwestern wall a steep face with moderately battered fault scarps. The west and south sides are much more irregular than the northeastern side, and the valley slopes are less abrupt. Most of the valleys are high on the southwest side, draining to the northeast as though the fault blocks have been tilted to the southwest (Hinds, 1952, p. 107).

**Photo 8 - Blair Valley. A view from S2.**

These faults have compelled the streams in most of the valleys, after reaching their northeast side, to flow southeastward, e.g., the Coyote and Grapevine Canyons. However, some streams, like the San Felipe and Benner, have eroded deep gorges cutting across the faults, suggesting that the streams were present before the fault began to develop and that movement along these fractures was so slow that the streams could cut downward as rapidly as the rocks were elevated across their path (Hinds, 1952, p. 107).

These faults are important to this study in that they have been the principal agents in the creation of the natural paths through which man moved from the interior to the coast. For instance, the path through Vallecito Valley to the Coyote Mountains became the route for the Overland Trail. The path through Carrizo Canyon became the route of the San Diego and Arizona Eastern Railroad.

Most of the sedimentary rocks were laid down as lacustrine deposits and alluvial fans, but included in the sequence are marine beds that probably accumulated during the Pliocene Period. This is indicated by certain sediments having marine fossils. The most fossiliferous and best-known marine formation, the Imperial, is evident and has been variously assigned Cretaceous, Lower Miocene, and Pliocene ages. According to available information, there appears to be 14,000 to 20,000 feet of sediments in the Salton Basin. Despite this thick section, the oldest formation, that of Split Mountain, does not appear to be older than the middle Miocene; a major part of the sediments being of Pliocene age, mostly non-marine in origin.

During Quaternary times, the Colorado silt-laden waters formed a huge delta, partially filling the depression and damming off its below sea-level northern portion from the waters of the Gulf of California. Adding to the effectiveness of the delta is a forty feet warp in the earth which lies underneath it, which may have been the casual factor in the formation of the delta itself.

During heavy floods, the wandering Colorado River on its broad and continually rising delta, instead of emptying into the gulf, would occasionally flow toward the west and then northward into the Salton Sink. In the recent geologic past, it did so for a long period of time, forming a freshwater lake approximately 100 miles long; 35 miles at its widest point, and over 300 feet in depth. It extended from above the present city of Indio, in Riverside County, to approximately 70 miles south of the present United States-Mexico boundary (Jaeger, 1957, p. 86). At one time, it occupied an area approximately 2000 square miles and was, in various parts, 30 to 40 feet above sea level, filling the basin to almost overflowing (Hartman, 1968, p. 126). This ancient lake is known as Lake Cahuilla, named after the nearby indigenous culture of that same name. This lake probably lasted from 1000 A.D. to 1500 A.D. when the main course of the Colorado River became oriented south toward the Gulf of California. The basin remained comparatively dry; there was left but a salt encrusted playa with only a small marsh at its center, kept moist by a few springs and rain from cloudbursts. Then came the great flood of 1905 which created the present Salton Sea (Hinds, 1952, p. 103).

Because of the geologic history of the valley and surrounding mountains, the present landforms over much of the basin are associated with Quaternary alluvial deposits. The streams flowing into the basin have built alluvial fans around its borders, and flashfloods have washed materials toward the lower part of the basin (Durrenberger, 1968, p. 19).

The alluvium bordering the mountains that enclosed the valley contains sand, gravel, and silt that decrease in size into the finer, lacustrine sediments found in the central portion of the valley. Much of this Quaternary alluvium is covered by a thin veneer of lacustrine sediments composed of silts, sand, and clay which was deposited by Lake Cahuilla in recent times (Van de Kamp, 1973, p. 827).

Aeolian deposits have also played an important role in the geologic history of the area. Windblown sand, still accumulating, covers extensive areas of these alluvial deposits, as in the middle and southwestern portion of the Imperial County. The effect of wind in the formation of the landscape will be covered in the section on Wind Erosion.

# CHAPTER II

## The Peninsular Ranges Province

The Peninsular Ranges Province, as a general description, is that part of the Southern California batholith which separates the Colorado Desert from the Pacific Ocean. It is an igneous intrusion, which consists mainly of granitoid rocks, and underlies young and old, metamorphic, and sedimentary rocks over large part of the area (Durrenberger, 1968, p. 20). Up to a million years ago, this region was a lowland. Since that time, active faulting has caused the older surface to be uplifted and broken up into several southeast-northwest trending blocks. The surface of these blocks has been further broken up into many secondary blocks by variously trending faults. Some of these blocks have been tilted and uplifted, others have remained in place. Most of these blocks stand at different levels in a well-defined plateau. Few blocks have been relatively downfaulted forming valleys. A considerable part of the surface has been altered by erosion, and it has been covered by deep residual soils (Miller, 1957, p. 217).

The Peninsular Ranges are generally divided into four major areas, but this study will concern itself only with the eastern portion of the Peninsular Ranges which merge with the Colorado Desert. It is composed of four protruding mountain ranges with a general northwest-southeast trend: Santa Rosa Mountains, primarily south of the Riverside County line; Vallecito and Fish Creek Mountains; Tierra Blanca and Coyote Mountains; and In-Ko-Pah and Jacumba Mountains. The Laguna Mountains will only be mentioned briefly since they are mostly outside the area under study.

Between each of these ranges are intermontane corridors or valleys, which physiographically are considered extensions of the Salton Trough, and will be described in this section because their features are highly modified by these mountains. Each of these three corridors has one or more washes or creeks flowing into the Salton Basin.

The Santa Rosa Mountains
This portion of the Peninsular Ranges is located west of the San Jacinto fault, on the northeastern portion of the ranges. They comprise a tilted-fault block with a steep western front, thousands of feet high, and a much longer, more gradual eastern downslope reaching sea level (Miller, 1957, p. 219). The crest of the mountains is somewhat flat, providing miniature tablelands. The steep slopes have been carved by torrential streams creating deep, narrow canyons and have deposited extensive alluvial soils. The foothills along its southwestern base are composed of soft sediments which form a landscape best known as badlands (Hartman, 1968, pp. 93-96). Rabbit Peak, near the San Diego-Riverside County line, at 6623 feet above sea level, is the highest peak within this range. Coyote Mountain is separated from the Santa Rosa Mountains by Clark Valley. It has an elevation of 3192 feet above sea level and forms the northeastern end of Coyote Canyon extending as a broken ridge toward Borrego Valley. Coyote Mountain is almost devoid of vegetation, with deep ravines and valleys and unexpected playas and bolsons. It has an average slope of 29 degrees

at its base and diminishes at 1000 and 1600 feet above sea level. Its original gradient is regained at approximately 2000 feet above sea level, giving this portion of the peak the appearance of a plateau.

Vallecito and Fish Creek Mountains
Characteristic of these parts of the Peninsular Ranges are the badlands, flats, and intermontane mesas. The Vallecito Mountains are in the northern portion of the region, forming the entrance to Harper Canyon. Approximately 2.4 miles into the canyon is Harper Flat which has an area of approximately one square mile and a south-north running gradient of 3.6%. Separated from Harper Flat by 400 feet wide and a half mile wide ridge is Hapaha Flat. Harper and Hapaha Flats are both found at 2400 feet elevation. Hapaha Flat slopes from north to south at a gradient twice that of Harper Flat. This location marks the beginning of Fish Creek Wash.

A distinct topographic feature of the area is Split Rock Mountain, considered a conglomerate of geological patterns. There are granite and lava dikes, marine and freshwater fossils, anticline and syncline fault zones, landslides, igneous, sedimentary, and metamorphic rocks, and sandstone concretions. The whole mass is tilted and eroded into a labyrinth of canyons, washes, and broad valleys. See photo 9.

**Photo 9 - A view of an anticline on Split Mountain.**

Following the Fish Creek Wash, approximately eight miles south, a quadrisect mesa may be found. The middle and west sections of the mesa stand approximately 200 feet above the Vallecito and Carrizo Creeks. The eastern and south sections of the mesa are 200 to 300 feet lower than the middle and west sections. The eastern portion of the mesa is the farthest protruding portion of the peninsula itself, culminating in the Fish Creek Mountains, which tower dramatically over the desert floor with a slope of 33 to 38 degrees.

Tierra Blanca and Coyote Mountains

The Coyote Mountains are located between the Jacumba and the Fish Creek Mountains. This elongated northwest trending highland rises abruptly above the surrounding valley areas to an elevation of approximately 2400 feet. The range is approximately 50 square miles: 10 miles long and up to 5 miles wide. In profile, the Coyotes are asymmetrical due to geologic uplifting along the Elsinore fault and are tilted northward. The mountains consist of a variety of landforms and terrains ranging from broad alluvial fans at the lower elevations to the badlands on the immediate slopes, with precipitous and rough relief at the higher elevations. Bedrock forms a cove or central portion of the mountains and consists of various types and ages of rocks that have weathered differently resulting in the typical desert mountain terrain.

On the south side of the Coyote Mountains is Fossil Canyon where various fossil deposits have been explored. The gravel pits at the end of the canyon are very popular recreational areas for rock hounds. The range is composed mainly of granitic formations and marble. In the higher elevations, erosion has left reefs and hills of hard sandstone and fossiliferous formations which are consolidated with deposits of organic calcite. Talus slopes and cliffs are broken by occasional benches and precipitous ravines which carry water after infrequent rains. Erosion of the more resistant formations has created narrow, steep canyons with flat, sandy floors. In the lower areas, exposures of Andesite lava are overlain by a series of pastel colored siltstone, sandstone, and clay. The name Painted Canyon, for one of its most prominent canyons, reflects the coloring of the formations.

The Andesite consists of thick beds of tuff, breccia, and volcanic flow with colors ranging from browns to purples with occasional green and reddish tints. The Imperial Formation* crops near the end of the canyon and is characterized by rounded hills of silty clay beds with distinct brown and yellow colors. Intermittent stream flow has incised the rock, carving 100-foot-high vertical cliffs along the narrow channel. The channel bottom is flat and varies from about 25 to 75 feet wide. Alluvium from the watershed mantles the floor. A jeep trail follows along the canyon bottom to an elevation of 900 feet. The canyon walls are rough. Differential weathering has resulted in a series of small cliffs interrupted by shallow benches, creating a stair-step effect. Some stream undercutting along the ravine bottom is evident and vegetation is sparse throughout the canyon (Creole Corporation, 1980, pp. 2-3 to 2-9). Carrizo Peak, located on the northern portion of the range, is the highest point at 2480 feet above sea level.

The bedrock is exposed continuously in this area with only minor accumulations of slopewash, talus, and alluvium. Talus has accumulated on the shallow slopes or at the bottom of steep rock outcrops, while minor coarse gravel alluvium occupies some of the major drainages. Slopes surrounding the peak are usually very steep with occasional vertical faces several feet high. Complex folding or faulting in the immediate area has created rugged linear scarps and ridge features. A typical fault scarp is located along the western side of the Carrizo Mountains which is vertical and over 1000 feet high (Creole Corporation, 1980, pp. 2-1 to 2-2).

The dissection of some of these Tertiary formations has produced typical badlands terrain marked by numerous erosion channels. The Carrizo Badlands, northwest of Carrizo Peak, rise approximately 600 to 1000 feet above sea level. See photo 10.

**Photo 10 - A view of the Carrizo Badlands from County Highway S2.**

The southern grade of the Coyote Mountains has a 16-degree slope, whereas the east section, which borders the basin floor, has slopes at nearly 45 degrees. In the Imperial Formation of the Carrizo Peak, approximately three miles east of the Mica Road, sandstone formations called "chimneys" are found, most likely resulting from the infiltration of calcium carbonate.

In-Kop-Pah and Jacumba Mountains
This southernmost protrusion of the Peninsular Ranges is the most dissected of the four. For the most part, it is an extension of the Jacumba Mountains and is divided by the Carrizo Gorge, the In-Ko-Pah Gorge, and the Davies Valley. Typical of this peninsula are piles of boulders surrounded by only more boulders. These rock piles have gone through much exfoliation; the decomposed rocks being carried away to the lowest levels through canyons such as the Coyote and Palm Canyons, as well as many washes. See photo 11.

*Imperial Formation is of late Miocene to middle Pliocene times, and contains fossil oyster shell beds, mollusks, and some corals. It is composed primarily of interbedded sandstone and siltstone with scattered thin layers of gypsum. They are from 1 to 6 feet in diameter and approximately 5 feet in height. Marine fossils were found here (Childers, N.D., p. 1).

Photo 11 - View of desert landscape from S2. Jacumba Mountains in the background.

Table Mountain, a ridge-like extension of the Jacumba Mountains, at 4040 feet above sea level, and Mount Tule, at 4647 feet above sea level, are the highest peaks of the Jacumba and In-Ko-Pah Mountains, respectively. McCains Plateau, west of Jacumba, has a remarkably preserved, nearly level peneplain surface approximately 12 miles wide and 4000 feet high (Miller, l957, p. 219).

Mention should be made of the several conspicuous lava-capped hills visible near Jacumba in the southeastern section of San Diego County. These lava caps, which are erosional remnants of a once more extensive lava field, rest sharply upon late Pleistocene formations (Miller, 1957, p. 220).

## Other Topographic Features

Additional physiographic features located within the Peninsular Ranges and the Colorado Desert regions are considered significant enough to deserve further description.

Borrego Badlands

These badlands consist of recent deposits of stream and lake origin which were uplifted through faulting and later eroded by water and wind. They are full of fossils from Pleistocene time, showing that mastodons, saber tooth tigers, horses, and large birds inhabited this area. Technically, badlands are an extremely eroded, labyrinth-like irregular landscape, the result of flash-flood erosion on weak rocks, which become so broken and decomposed that they turn into clay. In the eastern portion of the badlands, erosion has uncovered layers of mud from ancient lake deposits. The presence of iron deposits gives the landscape a typical display of pink, yellow, red, and green coloring. Within the washes, a variety of plants and animals could be found, but eventually they became sparse to the east where the mudsills are located. Mudsills do not hold enough water for plants and animals to prosper, and any water available is extremely alkaline (Lindsay, 1978, p. 116). See photo 12.

Borrego Mountain
This mountain is composed of granitic rock that was formed by uplifting thick sedimentary deposits. One granitic body, a pluton, intruded into these sediments from below. Later, a second pluton intruded into the first layer producing large cracks. These cracks were later filled with materials such as feldspar, quartz, and mica from the second pluton. These cracks can be seen today as varicolored dikes and veins in the mountain (Lindsay, 1978, p. 72).

**Photo 12 - A view of the badlands at the eastern edge of the Borrego**

Split Mountain
This mountain features a variety of unusual phenomena, beginning with the split itself, and including other features such as anticlinal folds, narrow chasms, fossilized oyster shells, perpendicular canyon walls rising more than 600 feet, mud palisades, perpendicular concretion holes, and soft clay lifeless hills. The gorge forming Split Mountain is an antecedent stream canyon (Lindsay, 1978, p. 64). See photo 13.

**Photo 13 - Split Mountain**

Carrizo Creek and Carrizo Corridor

The Carrizo Gap, through which the Carrizo Creek flows, is also known as the Carrizo Corridor. It is bounded to the north by the Fish Creek Mountains. Carrizo Creek begins in the Jacumba Valley approximately 15 miles southwest of Coyote Wells. While a larger portion of this valley lies on the Mexican side of the international border, its outlet is within the United States. The creek runs northwardly for approximately 18 miles where it joins the Vallecito Creek. It is later joined by the San Felipe Creek and shortly thereafter becomes a rather extended wash which ends in the Salton Basin. The higher hills within the valley have an elevation of approximately 400 feet above the present creek bed. The summits of these ridges have been diminished by erosion. See photo 14.

The corridor is laced with washes which remain dry for the greater portion of the year and become torrential when rain occurs. The main wash, the Carrizo Wash, is found at the end of the sedimentary deposits of the Carrizo Creek, near Bow Willows. The wash is broad and level with mesquite and willows along its bed. The Carrizo Gorge drops in elevation from 2800 feet to the desert floor at 950 feet. Due to this range in elevation, there is a great variety of animal and plant life, from chaparral on the upper elevations to typical lower desert vegetation. Some palm groves can be found at the start of the wash. The Alverson Canyon found in this area provides extraordinary marine deposits (Parker, 1963, p. 77). Some bighorn sheep and mule deer live in the upper reaches of this rugged terrain.

Dos Cabezas

This subregion is a bowl-like valley on the eastern slopes of the Jacumba Mountains, with huge boulders scattered throughout. The alluvial valley gradually slopes toward the Salton Trough. One point of entry into this valley is via Dolomite Road. The white limestone in this area was once mined for the manufacture of roof tiles. The San Diego and Arizona Eastern Railroad ran along the eastern edge of the valley, until the great flood of 1976 which completely dismantled the portion adjacent to the Coyote Creek Wash.

Truckhaven Trail

This route was started as a wagon trail by Borrego Valley residents in the 1930s. Having its start at the base of the Santa Rosa Mountains, it follows a course across the sand dike surrounding Clark Lake and eventually ends at the top of the Coyote Mountain (Parker, 1963,

**Photo 14 - Badlands on S2 near San Diego County line.**

p. 33). The trail goes through some of the most arid areas of both provinces. To the north of the trail, near Travertine Point, are high reddish-brown mountains denuded of vegetation and cut by rugged canyons. To the south are the Mud Hills of the Borrego Badlands. Well into the Colorado Desert Province, the trail splits in two. The main fork, S22 leads down the Arroyo Salado near Highway 86. Presently, S22 is paved from Borrego Springs to Salton City. See photo 15.

Clark Lake
This lake was named after the Clark brothers who developed it for watering their stock during the 1890s (Gudde, 1962, p. 62). It is located at the northeast corner of Coyote Mountain and to the west of the Santa Rosa Range. It is typically a dry lake of silty soil into which the waters of the Santa Rosa Mountains and the Borrego Badlands drain (Lindsay, 1973, p. 2).

**Photo 15 - Carrizo Creek and Wash.**

Borrego Sink
This is a true "sink" because the earth's surface sunk below the surrounding area. The waters of the Coyote Creek flow into it, as does the Palo Verde Wash when it has water. The sink is an area covered by alkali dust, held in check by salt grass and mesquite. The northern border of the sink is a maze of washes and dunes, forming part of the Borrego Badlands. The Borrego Sink Wash furnishes drainage not only for the sink proper, but also for the entire badlands to the north. Following the Palo Verde Wash, one encounters Borrego Mountain, a volcanic island pushed through deep sediments similar to Black Butte near Ocotillo Wells (Parker, 1963, pp. 30-32).

Upper Borrego Valley
One of the most interesting depressions of the Peninsular Ranges Province is the upper Borrego Valley. It is surrounded by, but not part of, the Anza-Borrego State Park. The valley was homesteaded after WWI, but most attempts failed, and a great deal of the land was divided into large parcels. After WWI, deep wells were drilled to provide water for

agriculture. The principal crops are table grapes, cotton, livestock feeds, and winter vegetables. Poor soils and high-water costs make agriculture an expensive industry (Lantis, 1977, p. 68).

Grapevine Canyon, Yaqui Wells, and the San Felipe-Vallecito Valleys
Grapevine Canyon was for many years the main-traveled road out of the Imperial Valley via Kane Springs, Harpers Well, San Felipe town site, and Little Borrego. The route continued up the narrows through the San Felipe Wash to Yaqui Wells, and then up Grapevine Canyon to Warner's Ranch. The Julian traffic followed the San Felipe Valley and then moved westward through the Banner Grade. Present Highway 78, running through Sentenac Canyon, has since made the old route obsolete. The beautiful high desert cienega is called San Felipe or Sentenac Cienega.

Approximately half a mile up the San Felipe Wash from the Tamarisk Grove Campground at the Anza-Borrego State Park is the famous old seep known as Yaqui Wells. Although its waters are saline, it continues to be an important watering spot for travelers and fauna. A large stand of ironwoods may also be found here.

The San Felipe-Vallecito Valleys form a large trough of great historical importance because it was one of the passes out of the valley for people coming through the Carrizo Corridor. Up the San Felipe Valley to the north is Warner Pass where the water of the creeks divide. Those flowing to the north end eventually in the Pacific Ocean; those to the south make their way into the Salton Sea (Parker, 1963, p. 59).

Anza-Borrego State Park
Established in 1933, this state park is perhaps the best remaining example of the natural desert (Lantis, 1977, p. 68). The 470,000-acre park includes striking rock formations, representative flora and fauna of the Colorado Desert, and a rich history. For more detailed information on the park, contact: Lindsay, Lowell, and Diana-The Anza-Borrego Desert Region Wilderness Press, Berkeley, CA, 1978

## Evolution of the Peninsular Ranges Province
The Peninsular Ranges are an immense fault block, elevated with an upward tilting on the eastern side. This block is divided into smaller units along subsidiary fractures, which either lie at high angles or run parallel to the boundary system. Most of these faults were active during most of the Cenozoic period. Each adjacent block has had a somewhat different landscape evolution due to the magnitude and nature of the movements affecting them (Jahns, 1954, p. 19).

The eastern scarp rises abruptly above the desert floor for approximately 1000 feet, after which the slope is reduced considerably. This points out a somewhat recent movement which elevated a block that had already advanced to a greater age of landscape evolution. This is also indicated by the fact that the scarp is carved by comparatively few deep canyons, having at their mouth only small alluvial fans (Hinds, 1952, p. 197). Although the scarp does not have the actual slope of the boundary fault system, it has not receded very far. At the southern edge of the Santa Rosa Mountains, this boundary scarp, with a slope of 16.5 degrees, is in most cases neither high nor steep. (Hinds, 1952, p. 197). Near the eastern scarp, there are detached and short summit ranges, mostly oriented east-west. Some run parallel to the ranges' trend. The Laguna Mountains is an example of this formation. It rises over 6000 feet, but the increase in elevation is not uniform or gradual because the range has been broken into separate units which have been uplifted independently from one another. Some of these units stand out as plateau areas, while others, less uplifted and usually smaller, form basins or valleys. Many of these dislocated remnants of a former continuous erosion surface have been minimally affected by the increase in the force of the streams following the uplifting. A good example of these less affected remnants may be appreciated in the comparatively large area of the McCains Plateau between Jacumba and La Posta Valley (Hinds, 1952, p. 197). Below the older erosion surface lays deep residual soils and weathered rocks which projects newer materials (Hinds, 1952, p. 200).

The somewhat recent uplift of the ranges is further proven by the slight erosion and ruggedness shown on the eastern scarp and by the narrowness and depth of the canyons running through it. Evidence of a major fracture system along the base on the Peninsular Ranges indicates, without doubt, that this scarp is the product of a tremendous dislocation which caused the sinking of the Salton Trough or Salton Graben and the elevation of the range.

On the southern edge of the ranges, a group of northwest trending mountainous ridges were formed by parallel mountain blocks and are separated by sharply sunken blocks like Clark Lake, Collins, and Borrego Valley. The Santa Rosa Mountains are the most important of these ridges. They have a long, straight scarp a few thousand feet high on the southwestern side, above which there is a slope (the old erosion surface) descending from 8000 to 6000 feet above sea level toward the northeast. Beyond is the great eastern scarp (Hinds, 1952, p. 200).

In other areas, the eastern scarp is not as simple. This indicates parallel faulting with distribution of dislocation along the different fractures creating a step-faulting topography (Hinds, 1952, p. 200).


Geology of the Peninsular Ranges
The inland parts of the province, "which are underlain chiefly by igneous and metamorphic rocks of pre-Cenozoic age" include among others the Laguna and Santa Rosa Mountains. These high masses are bounded by the Colorado Desert to the east and reach up to 6000 feet in altitude. "Most of these scarps are disposed in echelon and mark the sub parallel traces of major zones of faulting" (Jahns, 1954, p. 29).

This eastern scarp is strictly defined by a large, eroded surface descending abruptly from a more uniform upland landscape to the floor of the Salton Basin (Hinds, 1952, p. 19). This inland portion of the province, constituting part of the Southern California batholiths, is composed mostly of gabbroic to granitic plutonic rocks and is underlain chiefly by igneous, metasedimentary, and metavolcanic rocks of the Paleozoic and Mesozoic Ages. "The metamorphic and most of the igneous rocks antedate this batholith and form an extensive but mostly subordinate part of the crystalline terrene" (Johns, 1954, p. 31).

Marine strata along the northeastern edge of the province are represented primarily by Upper Tertiary clastic strata. Non-marine strata, chiefly lacustrine and fluviatile in origin, are found throughout the interior of the ranges where they are preserved as upper parts of low fault blocks. Exceedingly deep sections of clastic non-marine strata of probably Miocene time appear at the eastern edge of the province, particularly beneath andesitic volcanic rocks near Jacumba. The volcanic rocks are mainly shallow intrusives, flows, and pyroclastic accumulations of andesite and basaltic composition (Jahns, 1954, pp. 31 and 41).

The oldest exposed rocks in the area are mostly "sedimentary strata and subordinate interlayered volcanic rocks that have been mildly to severely metamorphosed" (Jahns, 1954, pp. 31). In general, they are present as inclusions of different sizes, pendants, and screens around plutonic masses, the remains of a former very extensive terrene that was deformed by volcanic invasion. This old terrene is evident in large areas of the Coyote and Santa Rosa Mountains, where "quartzite, crystalline limestone phyllite, hornblende, and mica schists, and quartz-feldspar and gneiss may reach an aggregate thickness of 22,000 feet" (Jahns, 1954, p. 31.)

The metasedimentary rocks, when they appear in the Laguna Mountains, are known as the Julian Schist. They consist mainly of quartz-mica schist and feldspathic to vitreous quartzite, with minor amphibolite, quartz-mica amphibole schist, metaconglomerate, and recrystallized limestone" (Jahns, 1954, p. 33). The various rock types, generally intergraded across and along the strike, appear to belong to a series of shallow water sediments.

At the eastern edge of the province there are many, sometimes quite large, plutonic masses of granodiorite and tonalite which antedate the formation of the batholith. Many of these rocks are extensively granulated and some of them show a well-defined gneissic structure. The Stonewall granodiorite, which appears in the Laguna Mountains and adjacent areas, is a good representative. The large plutons commonly are, in part, bordered by extensive areas of injection gneiss, contact breccias, and other kinds of magnetic rocks (Jahns, 1954, p. 35).

The Paleozoic rocks in the area have been moderately metamorphosed and belong to the abietic amphibolite facies and green schist facies. Among the more common minerals are epidote, garnet, hornblende, muscovite, and biotite. These rocks were probably metamorphosed before the deposition of the Mesozoic rocks which later were also metamorphosed (Jahns, 1954, p. 37).

A great majority of the rocks have been affected by contact metamorphism particularly those which occur as thin inclusions in the rocks of the batholith. The most widespread products are feldspathic gneisses, tektite, and other types of reconstituted limestone and foliated rocks which have garnet, amphibole, and pyroxene (Jahns, 1954, p. 37). Rocks of the Southern California batholith are overly complex formations which are formed by plutons or intrusive units whose outcrops range from several miles to a few hundred feet. Many of these units are directly juxtaposed along contacts not readily identifiable, while others are separated by curved or straight septa of older metamorphosed rocks (Jahns, 1954, p. 35).

Tubular inclusions are evident in many tonalities and granodiorites. Sometimes they are so abundant that the rocks look like flow-layered breccias. Some of the plutons have borders of hybrid gneiss which were formed when igneous materials invaded the old metamorphic rocks. Generally, these gneisses are less abundant than the hybrid rocks which are related to the plutonic formations that antedate the batholiths age (Jahns, 1954, p. 35).

**CHAPTER III**

# Geomorphology

Introduction
To understand the geomorphic processes of an arid environment, one must bear in mind that these regions have been subjected to recent violent orogenic forces, and to drastic climatic changes for at least the last half million years (Dunbier, 1968, p. 8).

The borders of the present desert areas, mountainsides, and hills may be mantled with talus, rubble, or soil; a condition inherited from an earlier age of semi-arid or even temperate climate (Blackwelder, 1954, p. 18). The beginning of a warmer age also marked the beginning of increased aridity and diminishing vegetation; consequently, the formation of a soil mantle was not only delayed but began to be carried away by floods and winds. Such erosion caused the formation of gullies on slopes that were formerly smooth. Even talus slopes were gradually carved by enlarging ravines, thus leaving only wedge-shaped remnants which were eventually carried away leaving only exposed bare rocks (Blackwelder, 1954, p. 18). Later, the return of a colder climate permitted the process to repeat itself, thereby filling gullies and ravines with soil and rubble and thus restoring the former smooth surfaces. An alternating process of creep and flood appears to be the reason for the well-known rubble strips on many hillsides at the edges of the modern desert (Blackwelder, 1954, p. 18).

After the last Ice Age began to subside, the glaciers retreated from most of the American continent. Then a period of heavy rains continued. Pourade describes it:

> The Pluvial Period, which ended 10,000 years ago, was followed in historical times by what Rogers has described as the Long Drought which lasted for approximately 3,000 years from 5,000 B.C. to 2,000 B.C. This time was followed, he thought, by a wet period of 1,500 years from 2,000 B.C. to 500 B.C., one of sufficient precipitation to again produce streams and lakes in the California Desert; followed by still another, but shorter drought of 1,000 years between 500 B.C. and 500 A.D. (Pourade, 1966, pp. 16-17).

Considering the above, it is evident that erosion has played a particularly important role, acting more rapidly against certain rocks and creating the present landscape of depositional plains, smooth carved rock pediments, and isolated rugged mountains. The greatly eroded mountain forms are mostly the result of long continued sub-aerial denudation, and the alluvium-filled valleys, the subsequent deposition (Dunbier, 1968, p. 8).

Diastrophism
There is clear evidence that the all-important features of the desert landscape were created mainly by orogenic movements involving both warping and faulting. Locally, the down-faulting Salton Trough and the striated faults which line the desert were all rough-hewn through comparatively recent diastrophism (Pourade, 1966, p. 11). However, the most easily identifiable diastrophic features are the low fault scarps that interrupt the basal slopes of the

surrounding mountain ranges (Blackwelder, 1954, p. 11). Most of these features occurred slowly, so that the relatively rapid erosion has greatly obscured or even totally changed those features produced before the late Pleistocene era.

Exogenous Geomorphic Processes
These processes are "all those physical and chemical changes which effect a modification of the earth's superficial form" (Thornbury, 1958, p. 34). Some of these processes which affect the desert landscape are weathering, erosion, and mass wasting.

*Weathering.* "Weathering may be defined as the disintegration or decomposition of rock in place" (Thornbury, 1958, p. 36). Rock deformation and fragmentation plays a particularly important role in the creation of desert features. Violent expansion and contraction due to rapid changes in extreme temperatures cause the rocks to fracture and comminute. Sheet-like structures in granitoid rocks is believed to be produced this way (Jahns, 1943, p. 67).

Another important process of weathering is hydration. It involves the expansion of constituent crystals, then the breakage of bonds between the mineral grains, and finally the disintegration into sand and rubble. In the initial stages of hydration, the expansion, no matter how slight, of the exposed parts of boulder or outcrop causes exfoliation. In this manner, joint blocks are changed into roundish boulders or "wool-sacks" forms typical of the granitic outcrops of the desert (Blackwelder, 1954, p. 12). In the desert, exfoliation results more from chemical weathering than from the effects of alternate heating and cooling of rock surfaces.

Weathering can be a preparatory process for erosion and makes erosion much easier, but it is not a prerequisite to, nor necessarily followed by erosion.

*Erosion.* "It is a comprehensive term applied to the various ways by which the mobile agencies obtain and remove rock debris" (Thornbury, 1958, p. 36). There are four aspects of erosion: the acquisition of loose materials by erosional agency, the wearing away of solid rock by impact upon it of materials in transit, the mutual wear of rock particles in transit through contact with each other, and transportation. The two main erosional agents are wind and water. Their work in the desert is extremely effective due mainly to the scarcity of vegetation.

Wind action is very evident on the desert landscape. In general, the desert is an area of deflation and suffers a net loss from its surface (Blackwelder, 1954, p. 14). By the process of saltation, the wind carries the coarser materials a few feet from the ground, the finer sand is carried much higher, and the dust is lifted hundreds of feet into the air, most of it carried away from the desert (Blackwelder, 1954, p. 14).

In depositing its load, the wind heaps fine gravel into windrows which often are a few feet or less in height. Sand is accumulated in dunes of varied sizes and shapes, or it is built into drifts in suitably placed ravines in the lee side of the obstacles. Such accumulation can be found generally on the lee side of an old lake bottom or in eddies localized by the surrounding mountains (Blackwelder, 1954, p. 14). The silt and still finer sediments form thin coverings of loess on the slopes of surrounding mountains and hills where hardy vegetation is scattered over the surface. The effectiveness of deflation depends on the height and spacing of the vegetation.

Deflation is a highly selective process because it sorts out the loose materials according to size and weight. From surfaces without vegetation, it removes all the finer materials. For instance, on the typical gravelly alluvium deposited by floods in a mountainous desert, wind action winnows out the finer materials, leaving only the small pebbles or coarser cobble elements. In time, a protective and continuous cover is formed, the typical desert pavement. It is an extremely hard, mosaic-like surface, a residual feature resulting from surface degradation of large expanses of mildly inclined alluvial materials. Rain and wind are the principal agents on its formation (Rodgers, 1966, p. 39). Since desert pavement takes hundreds of years to develop, it occurs only on upland plains and terraces which are free of frequent floods. This tessellated rock floor is common in the areas of igneous or volcanic rocks, but it can also occur in limestone tracks (Jaeger, 1956, p. 33). See photo 16.

Another side effect of deflation is the abrasive erosion caused by the winds. This erosion occurs mostly in places where surfaces are generally free of vegetation, where there is plenty of gravel, and where winds are strong. Thus, little evidence of abrasive erosion is found where shrubs are spaced 5 to 10 feet apart, including the ridges or hilltops and other areas far away from a supply of sand. Abrasive erosion is most effective on loose pebbles and stones lying on the desert plains, but such abrasion chisels the outcrops on the same plains (Blackwelder, 1954, p. 16). During wind erosion several stages are evident. First, the rock surfaces are barely pitted and rendered dull. Next, the surface develops shallow groves,

**Photo 16 - Desert pavement. A view from S22 in Imperial County.**

rounded corners, and incipient facets. In the last stages, abrasion develops distinctively shaped pebbles bounded by facets and rather sharp angles, such as the familiar dreikanter.

When conditions are suitable and the wind is parallel to the slope, the sand blast will erase preexisting, ridge-and-gully topography and leave a smooth slope. If the wind is parallel to a set of ravines and ridges, the former may hollow out into round bottom troughs leaving sharp ridges or fins called yardangs (Blackwelder, 1954, p. 16).

Much of this work is done by the everyday wind, but the frequent whirlwinds and the occasional gales have a greater cumulative effect over time.

Water is a more important geomorphic agent than wind. Even in the driest areas of the desert, stream erosion has played a vital role in shaping the desert landscape.

The typical desert stream is a flood that develops quickly after torrential rains and runs in otherwise dry channels seldom lasting more than a few hours (Rodgers, 1966, p. 12). Because of the sparsity of vegetation, the volume of water involved, as well as the gradient, waters run very rapidly creating a tremendous erosive power. The typical desert mountain, with its crenelate summits, projecting spurs, and major irregularities, has been produced, in part, by the differentiating effect of erosive streams. This, in turn, may lead to the formation of a pediment.

Perhaps even more effective than stream floods are desert sheetfloods. This term, coined by McGee (1897), refers to broad sheets of storm-born waters which move in a system of small channels, rather than in a definite stream course. Sheetfloods take place when there is torrential rainfall on a barren, detrital surface and there are no carved channels to prevent the spreading of sheetfloods. The striking features of sheetfloods are the brief duration, short distance of flow, and the fact that the water becomes almost immediately loaded with debris due to the presence of loose detrital material (Thornbury, 1958, p. 281).

The origin of desert basins is intimately related with that of mountain ranges. Even if the valleys are primarily erosional depressions, they are constantly being filled with detritus carried away from the mountain slopes. The geomorphic process in the formation of desert basins is that of transportation and deposition (Dunbier, 1968, p. 110). The alluvial plains of the desert basins may seem flat, but they raise steadily toward the marginal mountains, in this case, the Peninsular Ranges.

Rubble erosion, although minor, is still significant. It occurs more noticeably along steep slopes of unconsolidated gravel, such as the terraces that are being undercut by desert floods. The rolling pebbles excavate round-bottomed gullies and build evanescent talus cones below. Wind also dislocates pebbles on very steep rifts. Rubble erosion is also significant to the rill action of slopes on many barren mountains (Blackwelder, 1954, p. 14).

Rain pelting is also a minor but significant geomorphic process. In areas sheltered from the strong action of floods and winds, the soil surface is subjected to vertical erosion by pelting rain and hailstones.

The result is an abundance of small pillars, many of which are protected by tiny pebbles or lichen. The beating of rain also displaces loose particles and contributes to their slow but continuous downslope movement (Blackwelder, 1958, p. 14).

*Mass-wasting.* "Involves the bulk transfer of masses of rock debris down slopes under the direct influence of gravity." (Thornbury, 1958, p. 36). Rock falls, debris falls, rockslides, and subsidence are all examples of mass-wasting. It occurs when unconsolidated or weak materials become dislodged and tumble-down steep slopes.

Major Landforms of the Arid Regions
Three major mountain slopes exist within the area under study. In one case, the plains meet the mountains in a smooth curve which is concave upward. This is most likely caused by the action of rills and sheetfloods moving down the slopes and indicates that the mountain mass is stabilized and an older structure, e.g., Mount Signal. In the second case, the plains meet the mountainsides abruptly with a slope of 34 degrees or a gradient of 56%. This type of slope seems to exist only where the mountain has undergone conditions of rapid rise, e.g., Fish Creek Mountains (Blackwelder, 1954, p. 17). The third variety of gradient is a combination of the first two, indicating a compound condition and a complex history, e.g., Coyote Mountains.

Mountains in the arid regions are generally bordered by smooth piedmont slopes which extend to the basin floor. It was formerly thought that a piedmont was only composed of aggradational materials, but it actually consists of two parts: a lower portion of aggradational origin, called a bajada; and an upper portion which is an eroded rock surface generally covered with a mantle of alluvium approximately two to five feet thick (Sutton, 1966, p. 89). This bedrock portion is called a pediment.

Pediment and bajada slopes are relatively gentle, varying between one half degree and 7 degrees, whereas the mountain fronts are typically much steeper.

When normal stream gradients have been interrupted by faulting or other orogenic occurrences, the material washed down from the canyons is spread about the base of the mountains creating an alluvial fan as the stream tends to reestablish its normal gradient (Blackwelder, 1954, p. 17). When alluvial fans are distinct and the erosional form is visible, the term upper bajada is used. After fans have disappeared and have formed a gentle, uniform plain of aggradation, the term lower bajada is used (Dunbier, 1968, p. 12).

Lower bajadas seldom rise more than one or two degrees from the flood plains or desert playas at the bottom of the basin. The flood plains are located where the slope of one mountain meets the slope of another mountain; it is usually marked by a wash, which is considered the axial streamway of a desert basin (Dunbier, 1968, p. 12).

At times, pediments and alluvial fans are confused with each other since they have a general resemblance. However, pediments invariably develop where conditions are relatively undisruptive. Fans, on the other hand, are distinguished by a more complex profile and have a steeper gradient. An important point is that fans are symptomatic of the early stages of erosion, whereas pediments usually develop more as the erosion cycle proceeds

(Blackwelder, 1954, p. 17). An arroyo is a Spanish word used often to describe a sharply incised channel or wash. They are generally V-shaped in cross-section, although some of the larger ones have flat bottoms. The walls can be vertical or flare at various angles. An arroyo which begins in a desert mountain may continue along the upper bajadas, but if it should dwindle down, become shallow or reticulate, it may disappear altogether. If this occurs on an upper bajada then run-off or sheetflooding takes place; the heavier material will be deposited first, and the lighter material carried farther. The result is degradation of valley fill with coarser materials in the upper bajadas and progressively finer material on the lower bajadas. If a playa should exist, it will contain the finest materials. See photo 17.

**Photo 17 - An example of an arroyo.**

The existence of a playa will depend, not only on the ability of the run-off to reach the basin center, but also on the absence of an outlet to a drainage system (Dunbier, 1968, p. 12). When the water reaches the basin center it is usually charged with salts and other alkalae. Salt flats and selines will then occur. A playa may cover extensive areas, but it is never very deep. A playa basin may overflow with a single downpour, but generally will take several storms during the rainy season to create this temporary condition. Playa waters do not remain around long enough to allow waves to perform significant erosion along its edges. Because perennial lakes have occurred around Pleistocene times, it is common to find wave-eroded terraces, gravel bars, and other characteristic features around basins of modern desert playas and dried lakes. See photo 18.

**Photo 18 - Playa. East of San Felipe Wash.**

Although many dried water sites are underlaid by hard compressed clay and silt material, some are covered with a crust of salt. The difference is that in the clay playas, ground water finds it possible to slowly filter out of the basin to a lower place, or to the substratum; whereas in the saline basins, the only available exit for the disposal of ground water is by evaporation. The term cementation refers to the process where capillary water seeps just below the surface. When this happens, there is a tendency for loose materials to become cemented by carbonates and other salts, creating the well-known "caliche" or hardpan, which is the formation of a saline crust upon the surface of a hardening soil or outcrop.

Other Features of Arid Landforms
The term ground patina refers to a coating process, which does not necessitate an initial chemical reaction, and will form under rocks made of glossy materials which have rested in a permanent position. The coating is the result of capillary action, but its formation depends on the presence of ferruginous soils or acidic red soils (Rodgers, 1966, p. 33).

Desert varnish refers to the lustrous brown or black veneer consisting largely of oxides of iron and manganese found on exposed areas of rock (Blackwelder, 1954, p. 14). The origin of this process is not very well understood, but it seems clear that it is not an exudation of the rock, but rather a decomposition of algae, bacteria, lichen, and its spores, all of which have a high content of iron and manganese salts. Following their decomposition, the salts become oxidized through long exposures to intensive sunlight and air. After a time, this coating gains a high luster through the buffing action of the wind-driven silts and fine sand particles (Rodgers, 1966, p. 35). Desert varnish is a worldwide phenomenon and may occur at any altitude. In the area under study, it is mostly confined to the lower altitudes, and rarely found above the 1000 feet contour (Rodgers, 1966, p. 35).

Any kind of rock may be affected by desert varnish regardless of its porosity, but it is most often found on glossy materials such as jasper, obsidian, and chalcedony. These rocks, although of a highly siliceous nature, are particularly resistant to chemical reactions, so if there is an incipient development of desert varnish, it is because they have associated with very alkaline soils. Desert varnish generally occurs in association with desert pavement.

**CHAPTER IV**

## Climate

The climate of the Colorado Desert Province and the eastern areas of the Peninsular Ranges are characterized principally by extreme heat and low humidity.

Several factors contribute to the aridity of the area. Dunbier summarizes as follows:

(1) A location at the eastern margin of the North Pacific high where normal subsidence contributes to the stability of atmospheric conditions
(2) Additional subsidence along a coastline where surface air is controlled by a cold ocean current
(3) The relative weakness of any internal pressure system
(4) Distance from the polar and tropical zones of convergence
(5) The heating of the incoming air
(6) An intermontane location
(7) Intense local sunshine and evaporation (Dunbier, 1968, p. 15).

The study area falls almost in its entirety, in the lee shadow of the Transverse and Peninsular Ranges. These mountains, which surround the basin, take in most of the moisture that could be otherwise available to the lowlands. Most of the area under study falls within three climatic classifications: Low Desert, High Desert, and Steppe or Transitional climates.

Low Desert

The Low Desert climate is the driest, warmest, and sunniest of all California climates. Most of its stations record less than 5 inches of precipitation annually, the average January air temperature is approximately 54°F , the average July air temperature is approximately 92°F, and the sun shines approximately 90 percent of daylight hours. This climate corresponds to the modified Koppen's classification Bwhh. Table I gives data on temperature and precipitation by month as recorded at Imperial, California, for a period of approximately 50 years.

The average annual temperature is approximately 73°F. The lowest temperature recorded was 16°F on January 22, 1937. The highest temperature recorded, 119°F, happened on four dates, July 14 and 16, 1936, July 25, 1943, and June 25, 1970 (USDA, Soil Conservation Service, unpublished report, p. 3). The frost-free period for Imperial, California is more than 350 days in 3 out of 10 years, and more than 300 days in 9 out of 10 years. On average, there are approximately 8 days per year of frost in Imperial. Frost distribution, intensity, and frequency vary within the irrigated and non-irrigated lands. Data recorded at 30 different fruit frost temperature stations throughout the Imperial Valley show an average of approximately 32 nights per year when temperatures drop to 32°F at one or more stations (USDA, Soil Conservation Service, Unpublished Report, pp. 3-4). Average diurnal temperatures range between 20 to 30 degrees throughout the year.

Average annual precipitation is approximately 2.8 inches. Yearly rainfall averages, however, may be misleading as there is considerable variation from year to year. Yearly rainfall has varied from a fraction of an inch to over 8 inches. June is the driest month of the year, with measurable rainfall occurring only once, 0.04 inches on June 2, 1914. The rainy season is from August through March, during which there is an average of little more than 3 hours of rainfall per month. The highest rainfall for one day was recorded on September 6, 1939 when 4.08 inches of rain was recorded. September 1939 was the wettest month, with 7.06 inches and 1939 was the wettest year, with 8.52 inches. The only general snowfall ever recorded occurred on December 12, 1932, with 2.5 inches at Imperial, CA and 4 inches reported in the southeast portion of the Imperial Valley (USDA, Soil Conservation Service, Unpublished Report, p. 4).

Humidity is low, especially in the summer, but relative humidity has increased because of irrigation, and more recently with the filling of the Laguna Salada, in Baja California, Mexico, a short distance away. The average annual relative humidity is approximately 29 percent in the irrigated areas; the average monthly relative humidity reaches a maximum of 40 percent in August, and a minimum of 24 percent in March and April.

The intensity of aridity in the desert area is, perhaps, best described by what is known as the precipitation-evaporation ratio. This is the ratio between probable evaporation and actual precipitation; it also shows the discrepancy between the water upon which plants can draw and the desiccating effect of the desert sun heat (Jaeger, 1956, p. 43). Evaporation for the Salton Sea is 30 times its precipitation (Dunbier, 1968, p. 75).

Prevailing winds are westerly in the spring and winter. On windy days, velocities of 15 to 20 miles per hour are not uncommon, and gusts sometimes exceed 30 miles per hour, particularly where the desert floor joins the mountain escarpments to the west. The breezes in the summer months are generally less than 15 miles per hour, tend to come from the southeast, and bring humid air from the Gulf of California (USDA, Soil Conservation Bureau, Unpublished Report, pp. 4-5).

Temperature inversion layers forming during the night are prevalent throughout the year. The bases of these layers may be on or near the surface and extend as high as 600 feet. These inversions tend to break up during the early winter months.

A description of the seasonal weather variations for the region is given, following Bennet's analysis (Bennet, 1975).

During the winter months, the basin is normally covered by a somewhat intense anticyclonic circulation, except during the periods of frontal activity. At this time of the year, approximately 20 to 30 frontal systems find their way into the northern part of the desert basin. These systems are generally weak and become more diffuse as they move south into the Salton Basin. Most of the precipitation which occurs in the area is associated with these frontal systems. The Salton Basin is protected, in general, from the cold air masses which tend to move from Canada; this, together with the relative low latitude of the area, helps explain the large number of frost-free days during the year. Spring is a period of transition from the frontal activity to the very dry summers; temperatures rise, and precipitation decreases. During the summer months, the Pacific High is well developed, and a thermal trough overlies the southeast desert basin. The relative humidity is quite low during these months, decreasing to approximately 10 percent during the hottest part of the day. Temperatures over 100°F are common during the summer. Fall is again a period of transition. Temperatures decrease to milder levels, and precipitation again increases to an average of 0.3 inches per month.

High Desert
The low desert merges imperceptibly with the high desert as the terrain ascends gradually to the Peninsular Ranges. Interpolation based on the few places with climatic data permits the assumption that in the vicinity of the 1500-foot contour, the July mean decreases to 90 °F and the January mean falls below 50°F. This corresponds approximately to the modified Koppen's classification Bw. In areas where the annual precipitation drops below l0 inches, the climate has been designated as Low or High Desert climate, depending upon temperature conditions. Closer to the Peninsular Ranges, where the desert floor rises above the 3000 feet, the mean July temperature drops below 80°F. Despite the relative coolness in summer, the High Desert might seem oppressively warm on a summer day, and winter days will seem chilly due to the somewhat higher relative humidity. Spring and fall are pleasant. The wind is often strong; the lack of vegetation allowing active air movement much closer to the ground than in more humid climates where vegetation cover is more profuse.

Low relative humidity prevails in summer's daytime hours, and the air seldom feels damp during the year (Baily, 1966, pp. 40-14). Snow is common to the High Desert, and sometimes it covers the ground lightly for a few hours or even days. Roads above the altitude of 4000 feet  are sometimes clogged with snow. Under fair weather conditions, the desert receives the air from surrounding, higher slopes. Hence, frostier winter mornings may find a thin sheet of ice over standing water on low-lying ground when temperatures have remained above the freezing point on a nearby hill (Baily, 1966, p. 41). Table II shows climatic data for a station located in the High Desert climatic region.

Transitional or Steppe
In the interior valleys and mountain slopes, the climate is more typical of the steppe classification, Koeppen's modified Bs, which has less seasonal ranges in temperature.

The transitional climatic area lies in the broad ribbon along the interior faces of the Peninsular Ranges, midway between the valley floor and ridge crests, where the terrain is rough and steeply sloping.

The annual precipitation averages between 8 and 16 inches; the range of precipitation varies from 30 inches in the west to 6 inches in the east, but it is the areas enclosed by the 20-inch rainfall line that  stand out on the climatic map as islands of distinct mountain vegetation and climatic type. Surrounding these areas are the drier, and generally lower zones with the Transitional or Steppe type of climate. Midwinter (November through January) has the highest amount of rainfall with August recording the most precipitation for the summer months. Average monthly annual mean and maximum precipitation amounts are given on Table III.

The mean annual temperature for the area is 68°F. Average monthly temperature ranges from 44°F in January to 73.5°F in July. Table III shows monthly means and extreme temperatures recorded at a weather station located within the Transitional climate region (USDI. BLM. Proposed Livestock..., 1980, pp. 2-4). The daily relative humidity varies from 50 percent during winter months and 40 percent during the summer months.

The topography affects the length of the frost-free periods. The western part of the area has a frost-free period of approximately 150 days. The eastern portion has a frost-free period of approximately 200 days. Snow is a usual winter event at altitudes over 4000 feet and patches of snow last through the spring into early summer on high, shady northern slopes (Bailey, 1966, p. 34).

Running through the areas of Mountain and Transitional climates are the lines that denote 30-degree differences between the mean of the warmest and of the coldest months of the year. Sharper seasonal differences in temperatures, drier air, and a few clouds give the interior phase of the transitional climatic region a semiarid feature.

On the high summits, precipitation is more copious. Crests of the Laguna Mountains average 40 inches per year. Winters are relatively cold and total snowfall averages 3 to 4 feet per year. Summers, by contrast, are relatively warm (Lantis, 1977, p. 168).

TABLE I - TEMPERATURE AND PRECIPITATION, IMPERIAL, CA.

| Month | TEMPERATURE (DEGREES FAHRENHEIT) | | | | | PRECIPITATION (INCHES) | | |
|---|---|---|---|---|---|---|---|---|
| | Average Daily 1/ | Average Daily Maximum 2/ | Average Daily Minimum 2/ | 2 Years in 10 Will Have at Least 4 days With Max temp. Equal to or Higher than 2/ | With Min temp. Equal to or less than 2/ | Average Monthly Total 3/ | One Year in 10 Will have Less than | More than |
| January | 54 | 80 | 29 | 81 | 28 | 0.37 | 0.01 | 0.77 |
| February | 58 | 84 | 33 | 86 | 32 | 0.43 | 0.00 | 0.69 |
| March | 64 | 91 | 38 | 92 | 36 | 0.18 | 0.00 | 0.64 |
| April | 71 | 99 | 45 | 97 | 44 | 0.14 | 0.00 | 0.61 |
| May | 78 | 105 | 51 | 107 | 52 | 0.01 | 0.00 | 0.01 |
| June | 85 | 113 | 57 | 114 | 57 | 0.00 | 0.00 | 0.00 |
| July | 92 | 114 | 66 | 115 | 69 | 0.07 | 0.00 | 0.07 |
| August | 91 | 112 | 66 | 112 | 68 | 0.40 | 0.00 | 0.32 |
| September | 86 | 110 | 58 | 111 | 60 | 0.37 | 0.00 | 0.82 |
| October | 75 | 101 | 47 | 102 | 50 | 0.27 | 0.00 | 0.48 |
| November | 62 | 89 | 36 | 90 | 41 | 0.11 | 0.00 | 0.45 |
| December | 56 | 81 | 31 | 80 | 32 | 0.49 | 0.00 | 0.78 |
| YEAR | 73 Av. | - | - | 116 Max | 28 Min | 2.8 | 1.0 | 3.4 |

1/ Climatological Data, California, Annual Summary, 1974, U.S. Dept. of Commerce.

2/ 50 Year Summary, Records of Watermaster, Imperial Irrigation District.

3/ Climatological Data, California, Annual Summary, 1973, U.S. Dept. of Commerce.

From: USDA. Soil Conservation Service. Imperial Valley. Soil Survey Report.
    Unpublished Report 1962-1975, p. 56.

TABLE II – CLIMATIC DATA FOR TWENTY-NINE PALMS – HIGH DESERT CLIMATE

| Mean Temperatures – °F | | | | | | | | | | | | |
|---|---|---|---|---|---|---|---|---|---|---|---|---|
| Jan. | Feb. | March | April | May | June | July | Aug. | Sept. | Oct. | Nov. | Dec. | Year |
| 48.1 | 52.4 | 57.2 | 65.6 | 73.2 | 81.6 | 88.8 | 86.8 | 80.6 | 68.7 | 56.5 | 50.0 | 67.4 |

| Mean Rainfall – Inches | | | | | | | | | | | | |
|---|---|---|---|---|---|---|---|---|---|---|---|---|
| 0.6 | 0.3 | 0.3 | 0.1 | T | T | 0.5 | 0.7 | 0.4 | 0.5 | 0.3 | 0.5 | 4.2 |

| Temperatures | | | | | Relative Humidity | | Mean Annual Days – Sunrise to Sunset | | Mean Annual Snowfall (inches) |
|---|---|---|---|---|---|---|---|---|---|
| Daily Range | | | Extremes | | | | | | |
| Jan. °F | July °F | | Max. °F | Min. °F | Jan. | July | Clear | Cloudy or Partly Cloudy | |
| 26.7 | 32.6 | | 116 | 11 | 38 | 19 | 273 | 92 | 1.0 |

From: Baily, Harry P. *The Climate of Southern California*. University of California Press, 1966. pp. 85-87.

TABLE III – CLIMATIC DATA FOR WARNER SPRINGS – TRANSITIONAL CLIMATE

Precipitation – Monthly mean and extremes (in inches)

|  | Jan. | Feb. | March | April | May | June | July | Aug. | Sept. | Oct. | Nov. | Dec. | Year |
|---|---|---|---|---|---|---|---|---|---|---|---|---|---|
| Mean | 2.37 | 0.96 | 2.18 | 1.46 | 0.46 | 0.07 | 0.41 | 0.79 | 0.47 | 0.06 | 1.55 | 1.95 | 12.91 |
| Max. | 10.71 | 6.65 | 7.80 | 5.09 | 2.41 | 1.02 | 1.61 | 5.20 | 2.93 | 2.41 | 8.14 | 6.53 | ----- |

Temperatures – Monthly means and extremes (in $F^o$)

|  | Jan. | Feb. | March | April | May | June | July | Aug. | Sept. | Oct. | Nov. | Dec. |
|---|---|---|---|---|---|---|---|---|---|---|---|---|
| Mean | 44.4 | 46.2 | 48.2 | 52.2 | 58.3 | 65.0 | 25.5 | 73.0 | 68.2 | 59.5 | 50.9 | 45.5 |
| Mean Max. | 59.7 | 67.5 | 64.2 | 69.4 | 76.9 | 85.2 | 96.0 | 93.4 | 89.0 | 79.0 | 67.7 | 61.3 |
| Mean Min. | 28.9 | 29.8 | 32.4 | 35.1 | 44.8 | 53.0 | 53.0 | 52.5 | 47.5 | 40.0 | 34.0 | 29.6 |
| Max. | 84 | 84 | 89 | 92 | 102 | 106 | 110 | 107 | 105 | 97 | 88 | 84 |
| Min. | 11 | 14 | 17 | 19 | 25 | 27 | 34 | 28 | 28 | 22 | 35 | 9 |

From: USDI. BLM. *Proposed Livestock Grazing and Wilderness Management for the Eastern San Diego County Unit*, Riverside, California, 1980, p. 2-5.

**CHAPTER V**

## Soils

Introduction
Soils differ in their composition, productivity, appearance, and management requirements in different localities, or even within short distances. These differences are the result of interaction between the different soil-forming agents: (1) Parent material, (2) Climate, (3) Relief, (4) Time, and (5) Soil biotas (Thornbury, 1958, p. 74). The relative effect of these factors varies with each soil. In the desert, the climatic and landform factors take precedence over parent material. Most of the soils in the Colorado Desert area show little soil development because of the lack of moisture and vegetation. They tend to be alkaline because most of the moisture evaporates leaving dissolved salts behind.

The Salton Trough is an actively growing rift valley in which sedimentation has almost kept pace with tectonism. The basin has been partly filled by the accumulation of sediments carried by the water of the washes coming down from the Peninsular Ranges, but mainly by the silty sediments carried by the Colorado River since Miocene times. Formation of its delta perpendicular to the length of the Gulf of California rift has isolated the Salton Basin from the gulf, creating a close sedimentary basin. Since its formation, the Salton Trough has undergone cycles of filling with freshwater lakes and desiccation as the Colorado River changed courses, alternately, flowing north or south. In some areas, sediments attained a thickness of approximately 16,000 feet.

The eastern slopes of the Peninsular Ranges are steep, often greatly eroded, and their bases are commonly buried beneath upturned and eroded strata of conglomerate sandstone and clay, by masses of recent talus materials, or by the short but steep alluvial fans forming at the mouth of the small washes. Beyond these steeper basal slopes, the desert floor moves more gradually toward the lowest part of the basin, presenting a somewhat more uniform surface disturbed only by an occasional arroyo. Also, on the desert floor, are outcrops of igneous or sedimentary material which are now partially buried beneath the sedimentary fill.

Following the uplift of the mountain masses surrounding the basin area, there is evidence of an extensive period of erosion during which time there were deposited quantities of sand, gravel, and clay at the base of the ranges.

At a much later date, these sedimentary masses, now in the form of consolidated conglomerates, were subjected to a series of movements by which they were tilted and faulted. Following these periods of upheaval, these masses continued to be eroded, much resulting in the formation of badlands evident in the landscape today.

Another period followed in which the landforms were covered by a tremendous amount of coarser alluvial material, now exposed as the desert floor above the old beach line.

Coincident with this preposition of material by mountain streams and washes, the Colorado River deposited layers of silt, fine sand, and clay around the basin below sea level.

Since the last period of great deposition of materials around the mountain bases, there has been continual erosion by streams originating in the Peninsular Ranges. In narrow or somewhat restricted areas, the coarser materials have been completely removed, and the older and upturned sediments exposed more erosion. Examples of this may be seen on many of the lower slopes of the mountain ranges and in the vicinity of the Yuha Buttes.

 The more recent unconsolidated sedimentary materials represent the largest part of the desert soils above the old beach line along the eastern and western margins of the two areas under study. These more recent sediments form a gently rolling plain sloping toward the central portion of the basin. The entire surface is rather dendritic in nature (Strahorn, 1922, pp. 18-20).

The soils in this region are of various types, widely dispersed, and occurring at high frequency. Such a repetitive occurrence is due to the general terrain and the textural differences of the soils that develop in vertical sequences from mountain top to the floor of the basin. Exposure of the various soil types is due, in large part, to the faulting that occurred. In general, the texture of the upland soils is coarser, and that of the low-lying areas, thinner (Dunbier, 1968, p. 34).

Most desert soils are shallow with low humus content and lack a well-defined profile. Moisture is low and they lack the ability to retain water that is of utmost importance for vegetation. With such lack of moisture most of the year, harmful salts and carbonates are brought to the surface and precipitated by evaporation. On occasions, when an adequate amount of rainfall occurs, these minerals are carried off or leached out from the soils. In most instances, the drainage has a short distance to flow, and these minerals usually crystallize upon evaporation, leaving the characteristic hardpan or caliche (Dunbier, 1963, p. 34).

Soils of the Colorado Desert Province
There are 26 mapping units in the General Soil map for Imperial County. Of these, only 18 will be discussed in this paper, since the other 8 are out of the area under study. Most of these units are named for major soil series that occur within each unit. Soil series are groups of soils that have similar profiles or sequence of layers.

The soil series names are tentative and might be changed when the soils of the Imperial Valley are correlated with the National Soil Classification System. For now, detailed description of soils is on file at the El Centro office of the Soil Conservation Service (IID. Report for General Soil Map, 1967, p. 4).

The 26 mapping units are organized into 5 major groups based on soil characteristics and qualities.

<u>Group 1</u> - Areas dominated by fine-textured, noticeably young soils.

There is only one soil association belonging to this group in the area under study.

*Go-Im-AB2 Glendale-Imperial Association, 0 to 5% slopes, eroded*

The soils of this association occur on mostly level to gently sloping old lake beds and flood plains. They are developed in alluvium of mixed mineralogy. They are very deep, moderately-well to well drained, highly calcareous, and may contain gypsum and soluble salts. They are moderately eroded with numerous rills and gullies. Little of this association is cultivated or urbanized. This association is approximately 1.5% of Imperial County.

Glendale soils have light brown, subangular blocky, slightly hard, sticky silty clay loam or clay loam surface layers, and stratified light brown silty clay loam to heavy silt loam subsoils. Minor strata of both coarser and finer textures are frequently present in the subsoil. Subsoil permeability is moderately slow.

Imperial soils have light brown, subangular blocky, hard, sticky, plastic, silty clay surface layers, and stratified platy subsoils of materials like the surface. Most of the Imperial soils in this association contain more permeable strata at depths of 36 to 60 inches. Subsoil permeability is moderately slow.

Glendale soils make up approximately 40% and the Imperial soils approximately 40% of this association. Most of the remaining 20% is composed of inclusions of Meloland soils. There are minor inclusions of Gila and Vinton soils.

Group 2 - Areas dominated by medium and coarse-texture, young soils.

There are only five soil associations belonging to this group within the area under study.

*Gi-Vo-Awl Gila-Vinton Association, wet*

The soils of this association occur on nearly level flood plains, and old lake beds. They are developed in recent alluvium, or aeolian deposits of mixed mineralogy. They are very deep, calcareous, and may contain soluble salts. A drainage problem is present due to irrigation and high-water tables. There is a slight wind erosion hazard. This association is nearly all cultivated. It comprises approximately 9% of the irrigated area and approximately 2.5% of Imperial County.

Gila soils have light brown, subangular blocky, slightly hard, slightly plastic, silt loam to fine sandy loam surface layers and stratified subsoils like the surface layers. Subsoils permeability is moderately slow to moderate.

Vinton soils have light brown, massive, soft, fine sandy loam to loamy fine sand surface layers, with subsoils that resemble the surface layers. Subsoil permeability is moderate to moderately rapid.

Gila soils are approximately 40% and Vinton soils about 35% of this association. Approximately 15% of the association is composed of similar soils with major strata of fine textures in the subsoil. The remaining 10% is made up of inclusions which have fine surface texture and slowly permeable subsoils.

*Gi-Vo-A Gila Vinton Association, hummocky*

The soils of this association occur on nearly level flood plains, terraces, and old lake beds. They are developed in recent alluvium or aeolian deposits of mixed mineralogy. They are very deep, calcareous, and well drained.

There is a moderate wind erosion hazard. Almost none of this association is cultivated or urbanized. It makes up approximately 4% of Imperial County.

Gila soils have light brown, subangular blocky, slightly hard, slightly plastic loam surface layers, and brown, subangular blocky, slightly hard, fine sandy loam subsoils and substrata. Gypsum and soluble salts may be present. Subsoil permeability is moderate.

Vinton soils have light brown, massive, soft, fine sandy loam surface layers, subsoils, and substrata. Subsoil permeability is moderate to moderately rapid.

Gila soils are approximately 40% and Vinton soils about 35% of this association. Approximately 15% of the association is made up of soils with medium or coarse surface textures and control sections, but with significant fine texture strata within 60 inches of the surface. The remaining 10% is composed of inclusions of Imperial and Glendale soils.

*Mx-Gi-Awl  Meloland-Gila Association, wet*

The soils of this association are nearly level on flood plains and old lake beds. They are developed in recent alluvium and aeolian deposits of mixed mineralogy. They are very deep, calcareous, and may contain soluble salts. A drainage problem is present due to irrigation and high-water tables. There is a slight wind erosion hazard. This association is nearly all cultivated. It makes up approximately 13.5% of the irrigated area and approximately 4% of Imperial County.

Meloland soils have light brown, subangular blocky, slightly hard loam surface layers 26 inches thick. Thickness of the surface layer may range from 18 to 36 inches. The substratum is light brown, platy, hard, silty clay. Substratum permeability is slow.

Gila soils have light brown, subangular blocky, slightly hard, loam surface layers and brown, subangular blocky, slightly hard, fine sandy loam subsoils with moderate permeability. Most of the Gila soils in this association have fine textured substrata of slow permeability at depths of 40 to 60 inches.

Meloland soils make up approximately 55% of this association, Gila soils about 25%. Approximately 15% of the association is made up of inclusions of sandy soils with fine-textured substrata. About 5% of the association is composed of Imperial and Glendale soils.

*Mx-Gi-AB Meloland-Gila Association, hummocky, 0 to 5% slopes*

The soils of this association are gently sloping to nearly level on flood plains and old lake beds. They are developed in recent alluvium and aeolian deposits of mixed mineralogy. They are very deep, moderately well to somewhat poorly drained, calcareous, and may contain soluble salts. There is a slight wind erosion hazard. Almost none of this association is cultivated or urbanized. It makes up approximately 0.5% of Imperial County.

Meloland soils have light brown, subangular blocky, slightly hard loam surface layers 26 inches thick. Thickness of the surface layer may range from 18 to 36 inches. The substratum is light brown, platy, hard, silty clay. Substratum permeability is slow.

Gila soils have light brown, subangular blocky, slightly hard, loam surface layers and brown, subangular blocky, slightly hard, fine sandy loam subsoils with moderate permeability. Most of the Gila soils in this association have fine-textured substrata of slow permeability at depths of 40 to 60 inches.

Meloland soils make up approximately 55% of this association, Gila soils about 25%. Approximately 15% of the association is made up of inclusions of sandy soils with fine-textured substrata. About 5% of the association is composed of Imperial and Glendale soils.

*Vo-BM-AB1 Vinton-Brazito Association, hummocky, 0 to 5% slopes*

The soils of this association occur on gently sloping to nearly level flood plains and terraces. They are developed in recent alluvium and aeolian deposits of mixed mineralogy. They are very deep, somewhat excessively to well drained, and calcareous. There is a moderate wind erosion hazard. Almost none of this association is cultivated or urbanized. It makes up approximately 5.5% of Imperial County.

Vinton soils have light brown, massive, soft, fine sandy loam and loamy fine sand surface layers, and stratified subsoils like the surface layers. Minor lenses of fine-textured material are present in the subsoil in some places. Subsoil permeability is moderate to moderately rapid.

Brazito soils have light gray, single grain, loose, loamy sand or fine sand surface layers and subsoils of similar material. Minor lenses of fine-textured material are present in the subsoil in some places. Subsoil permeability is moderately rapid.

Vinton soils make up approximately 70% and Brazito soils about 20% of this association. Inclusions of soils with major fine-textured subsoil strata and Superstition and Acolita soils make up approximately 10% of this association.

<u>Group 3</u> - Areas dominated by coarse and very coarse, very young soils.

There are only three associations belonging to this group within our study area.

*Cr-Cd-A Carrizo-Cajon Association*

The soils of this association occur on nearly level areas of alluvial fans, basins, and flood plains. They are developed in recent alluvium of mixed mineralogy. They are very deep, somewhat excessively to excessively drained, and calcareous. There is a slight to moderate wind erosion hazard. Almost none of this association is cultivated or urbanized. It makes up approximately 9.5% of Imperial County.

Carrizo soils have very pale brown, massive, soft, loamy sand or gravelly coarse sand surface layers and gravelly sand subsoils. Permeability of the subsoil is rapid to very rapid.

Cajon soils have light brownish gray, single grain, loose fine sand or loamy sand surface layers and stratified subsoils like the surface layers. Strata of either coarser or finer textures sometimes occur at depths of 36 to 60 inches. Permeability of the soil is rapid.

Carrizo soils occupy approximately 50% and Cajon soils approximately 45% of this association. The remaining 5% is made up principally of inclusions of Rillito and Bitterspring soils, and soils with fine textured, slowly permeable subsoils.

*Cr-Cd-Bc Carrizo-Cajon Association, 2 to 9% slopes*

The soils of this association occur on gently to moderately sloping areas of alluvial fans, basins, and flood plains. They are developed in recent alluvium of mixed mineralogy. They are very deep, somewhat excessively drained, and calcareous. There is a slight to moderate wind and water erosion hazard. Almost none of this association is cultivated or urbanized. It makes up approximately 8% of Imperial County.

Carrizo soils have very pale brown, massive, soft, loamy sand or gravelly coarse sand surface layers and gravelly sand subsoils. Permeability of the subsoils is rapid to very rapid.

Cajon soils have light brownish gray, single grain, loose fine sand of loamy sand surface layers and stratified subsoils like the surface layers. Strata of either coarser or finer textures occur in some places at depths of 36 to 60 inches. Permeability of the subsoil is rapid.

Carrizo soils occupy approximately 45% and Cajon soils approximately 35% of this association. Approximately 15% of this association is composed of inclusions of Rillito and Bitterspring soils. Approximately 5% is made up of inclusions of soils with fine-textured, slowly permeable subsoils.

*N1-Im-AB2 Niland-Imperial Association, 0 to 5% slopes, eroded*

The soils of this association are nearly level to gently sloping on edges of old lake beds. They are developed in recent alluvium of mixed mineralogy. They are very deep, somewhat poorly drained, highly calcareous, and usually contain gypsum and soluble salts. They are moderately eroded with rills and gullies. Almost none of this association is cultivated or urbanized. This association is approximately 2.5% of Imperial County.

Niland soils have light brown, massive, slightly hard, gravelly loamy sand surface layers, and subsoils approximately 18 to 36 inches thick. The substrata are light brown, platy, hard, silty clays, with gypsum and soluble salts often present. Permeability is slow.

Imperial soils are light brown, platy, hard, silty clay throughout the profile. Only minor strata of more rapid permeability are found within 60 inches of the surface. Subsoil permeability is slow.

Niland soils have make up approximately 65% and Imperial soils approximately 30% of this mapping unit. Inclusions of Meloland soils and small narrow area of Carrizo soils make up the remaining 5%.

<u>Group 4</u> - Areas dominated by soils with lime segregation in the subsoil.

There are only two associations belonging to this group within the area under study.

*ss-Ab-A Superstition-Acolita Association, hummocky*

The soils of this association are nearly level on terraces, alluvial fans, and basins. They are developed in alluvium of mixed mineralogy. They are very deep, well drained, and have a slight to moderate wind erosion hazard. Almost none of this association is cultivated or urbanized. This association is approximately 8% of Imperial County.

Superstition soils have pinkish gray, massive, soft, calcareous fine sand surface layers, often with a thin gravel desert pavement. The subsoil is similar but with numerous lime concentrations in soft masses and nodules. Lime concentration diminishes in the deep substrata which is often clean sand at depths of 5 to 6 feet. Subsoil permeability is moderately rapid.

Acolita soils have reddish brown, slightly hard, calcareous, fine sandy loam surface layers, often with a slightly darkened gravel desert pavement. The subsoil is similar but massive and more compact and contains segregated lime in soft masses, threads, and/or nodules. Soft sand underlies the subsoil at approximately 40 inches. Subsoil permeability is moderately rapid.

Superstition soils make up approximately 55% and Acolita soils approximately 25% of this association. Inclusions of sand dunes, small basins with clay lenses in the subsoil, and small areas with plastic subsoils make up the remaining 20%.

*Rz-Os-AC Rillito-Orita Association, 0 to 9% slopes*

The soils of this association occur on nearly level to moderately sloping alluvial fans and terraces. They are developed in old alluvial materials of mixed mineralogy. They are very deep, will drained, and calcareous. There is a slight to moderate hazard of wind and gully erosion. Almost none of this association is cultivated or urbanized. It makes up approximately 2.5% of Imperial County.

Rillito soils have light brown, massive, slightly hard, slightly plastic gravelly sandy loam surface layers 3 to 8 inches thick. The subsoils are pink, massive, hard, gravelly loams to gravelly sandy loams with segregated lime in nodules, pendants, and discontinuous horizontal strata. Subsoils extend to depths of 50 inches or more and have moderately rapid to moderate permeability.

Orita soils have light brown, massive, soft, loamy fine sand surface layers 5 to 9 inches thick with light brown, massive, slightly hard, sandy loam, and loam subsoils up to 60 inches in depth. The subsoil has numerous lime nodules and is often moderately saline. Subsoil permeability is moderately slow.

Rillito soils make up approximately 65%, and Orita soils approximately 20% of the association. Inclusions of rough broken land, Carrizo, and Superstition soils make up the remaining 15%.

<u>Group 5</u> - Areas dominated by miscellaneous land types.

*AZ Alluvial Land Association*

This association occurs on low areas with a high-water table which is presently unfeasible to drain. There is no erosion hazard. Soil materials may range from clay to gravelly sand. Salinity ranges from slightly to strongly saline. This association often has a dense cover of phreatophytes. Almost none of this association is cultivated or irrigated. This association makes up approximately 1.5% of Imperial County.

Inclusions of higher, better drained land and rock outcrops make up approximately 10% of this association.

*BZ Badland Association*

This association is composed of highly dissected, rapidly eroding areas of unconsolidated material with little or no vegetative cover. Soil materials range from clay to loam in texture, with slow to moderately slow permeability. Almost none of this association has been urbanized, and none of it is cultivated. This association covers approximately 2.5% of Imperial County.

Approximately 95% of this association is badland, and 5% is made up of inclusions of deep, nearly level, sandy materials.

*DL Dune Land Association*

This association is composed of highly shifting sand dunes. The soil material is principally clean fine sand, somewhat excessively drained with moderately rapid permeability. Almost none of this association is cultivated or urbanized. It makes up approximately 4% of Imperial County.

Approximately 10% of this association is composed of areas of Superstition and Acolita soils between dunes.

*LP Lava Rock Land Association*

This association is composed of hills of basaltic materials with a complete cover of coarse gravel, cobbles, and boulders between rock outcrops. A calcareous soil material of sandy loam texture fills stone interstices to depths of 36 inches or more. There is no erosion hazard. Permeability of this material is estimated to be moderately rapid. None of this association is cultivated, and almost none is urbanized. This association covers approximately 1% of Imperial County.

Approximately 5% of this association is composed of inclusions of other types of rock land.

*PY Playa Association*

This association occurs in closed basins which are flooded with runoff water after rains. Soil materials are stratified silty clay, silt, and sandy loam. There are no inclusions of other soils. Water tables are not a limitation. Salinity is moderately high. There is no erosion hazard. Subsoil permeability is slow. None of this association is cultivated or urbanized. This association makes up less than 1% of Imperial County.

*RB Rough Broken Land Association*

This association is composed of highly dissected, rapidly eroding areas of unconsolidated materials of sandy loam and coarser textures. Erosion is extreme in high intensity rainstorms. Low intensity rainfall causes little erosion due to moderately rapid permeability of the materials. A fair cover of ephemeral plants will form following low intensity rains of sufficient volume. This association covers approximately 6% of Imperial County.

Approximately 15% of this association is composed of inclusions of Rillito, Bitterspring, and Cajon soils.

*RL Rock Land Association*

This association is composed of areas within 25 to 90% rock outcrop and very shallow soils. Principal rock types are granite and metasedimentary rocks. Most soil is eroded away as rapidly as it forms. Almost none of this association is cultivated or urbanized. This association makes up approximately 11% of Imperial County.

Inclusions of Carrizo and Cajon soils and shallow well-drained soils developed from the underlying rock make up approximately 20% of this association.

This information on soils for the Imperial County was extracted from the <u>Report for General Soil Map, Imperial County, California, 1967.</u> Soil Conservation Service.

Soils of the Eastern Portion of the Peninsular Range (Eastern San Diego County)
The soil associations in San Diego County have been assigned to eight groups. The grouping is based on soil qualities and characteristics and on their location within the province. Only three mayor associations fall within the study area, and they belong to three of the eight groups mentioned above. (For a more detailed description of the soils and their associations see the San Diego County Soil Survey, 1973, listed in the bibliography).

<u>Group 1</u> - Excessively drained to well drained, nearly level to moderately sloping very gravelly sands to silt loams on alluvial fans in desert areas.

*Mecca-Indio Association*

Well-drained sandy loams and silt loams on alluvial fans, subject to occasional flooding and deposition; 0-5% slopes.

This association is made up of soils that developed in alluvium derived from acid igneous rock and mica schist. It occurs in the desert.

The elevation ranges from 100 to 2500 feet. The mean annual precipitation is between 3 and 8 inches, and the mean annual air temperature between 70-74°F. The frost-free season is 220 to 275 days. The vegetation consists mostly of shrubs, cacti, and annual grasses.

Mecca soils make about 50% of the association, and Indio soils about 40%. Rositas and Carrizo soils and small areas of moderately to strongly saline soils make up the remaining 10%.

Mecca soils are brown and yellowish-brown course sandy loams, sandy loams, or fine sandy loams. Indio soils are pale-brown silt loam to a depth of approximately 45 inches, and below this pale-brown fine sandy loam. Both soils are slightly saline. Free water is seldom close enough to the surface to create a problem.

Irrigated areas are used for crops, most commonly cotton, dates, alfalfa, and small grains. Non-irrigated areas are used for range.

*Rositas-Carrizo Association*

Somewhat excessively drained and excessively drained loamy coarse sand to very gravelly sands on alluvial fans; 2 to 9% slopes.

This association is made up of soils that developed in alluvium derived from acidic igneous rock and mica schist. It occurs in the desert. The elevation ranges from 100 to 2000 feet. The mean annual precipitation is between 4 to 7 inches and the mean annual air temperature between 68-74°F. The frost-free season is 210 to 270 days. The vegetation consists mostly of desert shrubs, cacti, and annual grasses.

Rositas soils make up approximately 60% of the association, and Carrizo soils approximately 25%. Mecca soils and small areas of sand dunes make up the remaining 15%.

Rositas soils are somewhat excessively drained, light brownish- gray loamy coarse sands and fine sands that have a substratum of pale-brown, fine gravelly loamy coarse sand. Carrizo soils are excessively drained, very pale brown very gravelly sands and fine sands. They have a substratum of very pale brown very gravelly coarse sand.

Irrigated areas of Rositas soils are used in citrus, alfalfa, and pasture. Non-irrigated areas are used for ranges. Carrizo soils are too coarse texture for irrigated farming, but they provide a good source of sand and gravel for construction purposes.

Group 2 – Excessively drained to well drained, gently sloping to strongly sloping loamy coarse sands to sandy loams on alluvial fans and in basins in mountainous areas.

*La Posta-Kitchen Creek Association, rocky, eroded*

Somewhat excessively drained loamy coarse sands over decomposed granodiorite; 5 to 15% slopes.

This association is made up of soils that developed in material derived from granitic rock. It occurs on uplands in the mountains. The elevation ranges from 200 to 4500 feet. The mean annual precipitation is between 12 and 20 inches, and the mean air temperature between 56-58°F. The frost-free season is 160 to 190 days. The vegetation consists of chaparral, mainly chamise, red shank, scrub oak, and flattop buckwheat.

La Posta soils make up approximately 70% of the association and Kitchen Creek about 20%. Mottsville and Tollhouse soils make up the remaining 10%.

La Posta soils have a surface layer of grayish-brown loamy coarse sand and a substratum of brown loamy coarse sand. Below this is decomposed granodiorite. Rocky outcrops and boulders cover 2 to 20% of the surface. Kitchen Creek soils have a surface layer of dark brown loamy coarse sand and a subsoil of pale-brown coarse sandy loam. Below this is decomposed granodiorite.

These soils are used mainly for range, wildlife habitat, and watershed. Larger areas would be suitable for farming or for housing developments if water could be made available.

*Tollhouse - La Posta Rock Land Association, eroded*

Excessively drained and somewhat excessively drained coarse sandy loams and loamy coarse sands over granitic rock, and areas of rock land; 9 to 65% slopes.

This association is made up of soils that developed in materials derived from decomposed granodiorite. It occurs on uplands in the mountains.

The elevation ranges from 2000 to 5000 feet. The mean annual precipitation is between 15 to 20 inches, and the mean annual air temperature between 56-58°F. The frost-free season is 140 to 190 days. The vegetation is mainly chaparral and a few annual grasses and forbs.

Tollhouse soils make up approximately 45% of the association; La Posta soils about 35% and acid igneous rock land approximately 10%.

Tollhouse soils are excessively drained. They have a surface layer of dark grayish-brown course sandy loam that is underlain by hard granitic rock. La Posta soils are somewhat excessively drained. They have a surface layer of grayish-brown loamy coarse sand and substratum of brown loamy coarse sand that is underlain by weathered granodiorite. Rock land consists of areas were 50 to 90% of the surface is covered with boulders and outcrops of acid igneous rock. Very shallow soil material occurs in pockets between the rocks.

The soils of this association are used mainly for range, watershed, and wildlife habitat.

*Sheephead, Rocky Bancas Association*

Well-drained cobbly fine sandy loams over fractured mica schist; 9-65% slopes.

This association is made up of soils that developed in material derived from mica schist. It occurs in the mountains. The elevation ranges from 3000 to 6000 feet. The mean annual precipitation is between 20 and 30 inches, and the mean annual air temperature between

56-58°F. The frost-free season is 160 to 185 days. The vegetation is chaparral, including chamise and Manzanita, and a few annual grasses.

Sheephead soils make up approximately 70% of the association. Tollhouse and Holland soils make up the remaining 30%. Sheephead soils have a surface layer of dark grayish-brown cobbly fine sandy loam. Below this is fractured mica schist. Rocky outcrop covers 2-10% of the surface.

These soils are used mainly for range, watershed, and wildlife habitat.

<u>Group 3</u> - Miscellaneous land types of the desert, mountains, foothills, and coastal plains.

*Rock Land Association*

Dominantly exposed bedrock and large boulders. This association is 50 to 90% exposed bedrock and large boulders. It occurs in the desert, in the mountains, and in the foothills. The outcrops and boulders are chiefly granodiorite, quartz diorite, gabbro, mica schist, metavolcanic rock, and metasedimentary rock. A thin mantle of soils occurs in pockets between the rock outcrops.

This association occupies more than 49% of the Eastern San Diego County Area. Acid igneous rock land makes up 75% of the association. Metamorphic rock land and other rocky soils make up the remaining 25%. Rockland has no value for farming or ranching. It is used mainly for watershed and wildlife habitat.

This information on soils was taken from <u>Proposed Livestock Grazing and Wilderness Management for the Eastern San Diego County Planning Unit.</u> Draft environmental Impact Statement. USDI. Bureau of Land Management, Riverside, California, 1980.

# CHAPTER VI

## Biogeography

Introduction
There is no doubt that the total environment of the arid areas finds its greatest expression in the field of biogeography because of the number of species it supports and their unique manners of adaptation.

This total environment includes landforms, climate, soils, flora, and fauna, relating to and modifying each other in such a way that anything short of a holistic interpretation would be inadequate. The fragility of this interdependence is nowhere as critical as in the arid lands; any minor change in any of the elements can unbalance the system producing disastrous results for any of its components.

The "denizens of the desert" as Dr. E. Jaeger has so aptly called them, include both plants and animals marvelously adapted to the desert life (Jaeger, 1957, p. 10). They live harmoniously in defined groups called communities. Each community is made up of a population of individuals. Combination of communities in turn comprise biomes. In other words, in any given area, certain plants and animals have a special symbiotic relationship with each other. For instance, along a wash bed there may be a population of cottonwoods or willows (each tree being an individual) and associated with them may be a population of kangaroo rats.* In such a case, both populations depend on each other in some respect - the rats may disperse the seeds of the trees, while the trees, in return, may provide nourishment to the animals. (See appendix A for description of typical desert habitats).

Some habitats are isolated ecosystems such as oases, sand dunes, and springs. Like islands, each of these special habitats support a distinct animal population. Though rare, the wetlands such as marshes, oases, and riparian areas support a much denser population, as well as many species which only occur in those areas. The sand dunes, on the other hand, provide habitat for vertebrates found in other areas, but more importantly, provide a habitat for many invertebrates found nowhere else in the world (USDI-BLM, The California Desert..., 1980, p. 135).

San Sebastian Marsh, for instance, is one of those rare, perennial surface water systems in the desert. It is fed by the San Sebastian Creek. Here can be found one of the most important populations of the San Sebastian leopard frogs, which have their breeding ground at Harpers Well. The marsh also supports a large variety of other wildlife including 93 species of birds (USDI- BLM, The California Desert..., 1980, pp. 409-410).

*For genetic names of typical flora and fauna see appendix B.

Flora of the Colorado Desert

One of the major functions of vegetation is to serve as protector for soil, preventing erosion by binding the soil, sediments, and debris together. On the other hand, soils give the necessary support to sustain plant life.

Despite the harsh environment, the Colorado Desert supports a great variety of plants and animals. Besides its diversity, much of its plant life is very different from that of other deserts - so much so that the term "Sonoran" is used to distinguish the flora from the Colorado Desert from that of the other California Deserts (Lantis, 1977, p. 60).

One way that the flora and fauna can cope with a challenging desert environment is through specialization. Plants may be classified into three principal groups: Hydrophytes (water loving), Mesophytes (medium water use), and Xerophytes (drought resistant). Desert plants can also be either Halophytes (salt tolerant) or Glycophytes (low salt tolerant) depending on the salinity of the environment they can survive in (Jensen, 1972, p. 723).

For the most part, desert plants are xerophytic; that is, they can withstand drought through special means of water retention and conservation (Walton, 1969, p. 83). What is at first striking on the desert landscape is the spacing between certain shrubs. It is uniform as if following a plan, but root competition for water has resulted in a natural thinning process where competition favors the stronger individual plant.

Some physical adaptations of xerophytes to the arid environment may be noted in the succulents such as the agave which stores water in its fleshy tissue. The thin-walled cells shrink when water is scarce and swell up when water is plentiful. Their thick epidermis and waxy cuticle enhance the viability of this mechanism (Jensen, 1972, p. 723). See photo 19.

**Photo 19 - Desert agave. On S2 west of Ocotillo.**

All members of the agave and cactus family belong in this category. Members of these families are dominant in the vegetational composition in parts of the Colorado Desert at elevations between 3000 to 4000 feet. Of importance are the prickly pear, beavertail cactus, the yucca, and the desert agave. Teddy bear cholla frequently grows in single stands. Succulents are generally dispersed throughout shrub communities in the desert (USDI - BLM, The California Desert..., 1980, p. 188). See photos 20 and 21.

Photo 20 - Teddy bear cholla on S2 west of Ocotillo.

Photo 21 - Beavertail cactus.

Deep-rooted shrubs of the large desert streamways are also drought resistant and perennial in nature, e.g., mesquite, palo verde, ironwood, and smoke tree. Their deep and extensive root systems, reaching depths of as much as 175 feet, preserve their existence (Jensen, 1972, p. 723). See photo 22.

**Photo 22 - Wash with stands of smoke trees and other vegetation.**

The desert may be as colorful as a garden or as drab as a paved highway, depending on the moisture available. Not all spring seasons have spectacular wildflower displays. Rain must be of a certain quantity and arrive at specific intervals. In other words, one cloud burst in September bringing four inches of rain does not equal four showers bringing one inch of rain over a period of a month. While the first condition may be good for one species, the second combination of rains might destroy the same species.

The most common flowery plants display in the desert after adequate rains are the ephemerals or annual plants. These short-lived plants carry adaptation to an advanced degree. The seeds of these plants have a relatively long period of fertility. The natural dryness of the desert may allow a seed to lie dormant for years without degenerating. When and if rains occur, most but not all the seeds will germinate and grow into tiny plants. Regardless of weather fluctuations, their behavior and characteristics imply plenty of moisture at a given time. These plants tend to be emophytes.

At the western edge of the Colorado Desert Province, the Anza-Borrego Desert Park, a desert-within-a-desert, is a tangle of much eroded terrain, canyons, and badlands, complicating the eastern base of the Peninsular Ranges to the west. In this park, one may find plants that would be suspected of surviving only in northern Mexico or southeastern Arizona. The Anza-Borrego country has a claim to the fairy duster with its delicate puffs, and an arboreal oddity such as the elephant tree, one of very few representative species of the tropical Torchwood family. Like many desert trees, this one is peculiar in that the branches taper off rapidly giving the appearance of an elephant trunk. The trunk of the tree is swollen as well, and the stems are frequently leafless. These groves, and particularly those of the Carrizo Creek, are the largest in the United States (Parker, 1963, p. 92). See photo 23.

**Photo 23 - Elephant tree on the Anza-Borrego Desert State Park.**

Several other shrubs (which are mostly found in the Arizona desert) also appear in this region. They include the gray horn, California snakeweed, and the crucifixion thorn. The latter, a woody, thorny shrub approximately six feet tall, is one of the plants on the critically sensitive list for the area. It can be found in large stands near Mount Signal and in the vicinity of Ocotillo. See photo 24.

**Photo 24 - Crucifixion thorn. Stands near Mt. Signal.**

The sand dunes and sandy areas (the least favorable of all environments for plant growth) when stable, support the coarse perennial grass known as galleta grass. Moving dunes sometimes support the large, shrubby desert buckwheat, but otherwise, they are mostly devoid of vegetation. Such shrubs as the desert buckwheat, mesquite, creosote, or burroweed may give necessary resistance to wind-blown sand to start sand piling at the base - the start of a new sand dune.

Perhaps, the best show of color in the desert is made by the golden encelias and desert primroses, as well as the pink verbenas. Where the desert pavement is encountered, the show ends abruptly and only a few annuals such as the sand mat, the creosote bush, and the dropped blossoms of the ocotillo liven the mosaic-like surface. See photo 25.

**Photo 25 - Lone ocotillo at twilight.**

Upon the extensive desert plains, the vegetation is usually made up of low and open stands of creosote bush and burroweed. These two prevalences make these plants dominant species (Dunbier, 1968, p. 48). See photo 26.

Recent studies have indicated that creosote bush grows in rings, and that everything in such ring appears to have originated from a central area within the ring. As the center of the bush dies, new plants grow on the periphery. The low rate of growth for the ring and the size of the ring indicate that "some rings may date back to the end of the last Ice Age when creosote bush first established itself in the desert" (USDI-BLM, The California Desert..., p. 6).

The introduction of new species, particularly small trees, is common among the larger desert arroyos where there is considerable subdrainage. The size and distribution of drainage is important for vegetation. A dendritic pattern results in a very irregular spread of the larger species along the dry banks of arroyos. This lends a contrast to the sparse vegetation of the drier plains (Dunbier, 1968, p. 48).

**Photo 26 - Desert view with stands of creosote bush.**

When the drainage is reticulate, the mesophytic plants do not form a continuous border along the washes but are spread more uniformly over what appears as a plain or bajada community. On closer examination, however, one finds that their location and size are dependent on the location and size of less conspicuous reticulate drainage, even though the rainfall averages the same in a nearby district which supports more xerophytic vegetation.

On ascending the bajada, a gradual change in vegetation takes place; the frequency of plant species becomes greater, their size is larger, and new or different species are evident. This increase in vegetation indicates greater subterranean moisture and better root anchorage in the coarser soils. Ocotillo and catclaw are frequently found on the better-drained bajadas, as are chollas, barrel cactus, ironwood, and palo verde. See photos 27 and 28.

In most desert canyons, rim trails and gullies gradually meet to continue higher, leading often to one of the most striking features of the desert transitional landscape: lush oases, complete with springs and mesophytic vegetation. These vegetative outcrops not only have willows and cottonwoods but also Washingtonian palms, native to the region. This palm, remnant of a wetter era, maintains a "precarious membership in the flora of the Colorado Desert" (Jaeger, 1956, p. 189). The presence of these palms always indicates water, but not necessarily of good quality, for they tend to be halophytic in nature. The presence of arrowweed alongside the Washingtonian palms, on the other hand, indicates low salinity, for they are glycophytic. These canyons with their numerous crannies and overhangs give and adequate environment for a host of other plants such as phacelias, bladderpods, and four - o'clocks (Bakker, 1972, p. 266).

Photo 27 - Barrel cactus, hedgehog cactus and chollas on S2, west of Ocotillo.

Transitional Zone
The transition of species from desert floor vegetation to the chaparral and upper montane vegetation of the Peninsular Ranges takes place on the "transitional zone." This zone represents the typical vegetation of Southern California and occurs at approximately 3000 feet. The representative species of this zone are juniper, chamise, lupine, mormon tea, cholla, and barrel cactus, as well as other less dominant types. Vegetation of the transitional zone can survive the long drought of the summer months without difficulty.

This zone with its vegetation serves as a retardant to erosion and as watershed of the Chaparral region found above 3500 - 4000 feet.

Vegetation in this region grows densely, usually crowding out forage vegetation. In this region, fire is the agent of denudation. This occurs naturally approximately every 25 years. Primary succession takes place by the annuals, many only needing a little water, and by the "crowning" of certain perennials because of the excellence of their root stock.

**Photo 28 - Teddy bear cholla** and ocotillo on desert bajada.

Flora of the Peninsular Ranges

Above the transitional zone is the Chaparral, often a thick mass of evergreen shrubs, many times ten feet tall. It is able to resist long summer drought by means of adaptation; small or leathery leaves help plants conserve moisture. Distribution of the species varies locally; scrub oak is conspicuous at different habitats and elevations. At the lower elevations, chamise is frequently dominant. Farther upslope, manzanita and ceanothus become common. Chaparral is an important asset because it protects steep hillsides from surface erosion (Lantis, 1977, p. 86).

Within this community there are smaller subregions known as Oak Woodlands which might be described as an oak grove with grasslands. It generally occupies the north-facing hillsides and the intermontane valleys with small creeks. The Oak Woodland is dominated by the live oak, the valley oak, and the black walnut (Lantis, 1977, p. 87).

Within the Peninsular Ranges, at the higher elevations, are the larger timber areas of the Coniferous-Montane, where parklike stands of pine and cedar are present (Jahns, 1954, p. 29).

Fauna of the Colorado Desert and Peninsular Ranges Province
The fauna of the arid and semiarid environment depends upon plant life for survival. The abundance and type of vegetation, therefore, governs the number of species present.

In the desert and in the lower elevations of The Peninsular Ranges, there are two primary phases of plant growth: the lush period following adequate rainfall when most foodstuffs are abundant, and the drought period when most plants become dormant and food is scarce or unavailable. The fauna which survive in such an environment are those that have found ways to cope with the lean season (Leopold, 1961, p. 70). Some animals have adapted to the hot and dry season by avoiding them. The chuckwalla hibernates when food is scarce; the ground squirrel avoids the heat and thus retains water by sleeping in deep burrows below the scorching surface; some have nocturnal habits, such as the kit fox, which is a similar conservation technique; others are biologically equipped to conserve water, such as the desert tortoise which has actual water storage sacks as body organs (Lindsay, 1973, p. 17).

Vertebrates are more conspicuous and familiar but only make approximately 10% of the animal kingdom. The inconspicuous remainder may play a much larger role in the dynamics of a desert community (USDI-BLM, The California Desert..., 1980, p. 6). Slopes, peaks, and ridges with variable size rocks and boulders, and scattered vegetation provide den sites for larger mammals; the crevices are an excellent habitat for reptiles, but the sparseness of vegetation limits birds and small mammals (Creole, 1980, pp. 2-150-154).

In canyons with shallow ravines, water occurs more frequently and for longer periods than in the more elevated ridges and slopes; therefore, denser vegetation provides for better animal cover, and sand in canyon washes makes a good habitat for rodents. Narrow canyons and large boulders also provide cool shade for larger mammals (Creole, 1980, pp. 2-150-154).

Animals that may be glimpsed, particularly in the canyons away from the desert floor, are the coyotes and the magnificent bighorn sheep. Also, there are mule and black-tailed deer that wander over the arid mountains in the lean winter months, descending to lower altitudes during the drought season in search of springs (Jaeger, 1965, p. 11).

The black-tailed jackrabbit is probably the most frequently seen creature of the desert. The miniature ground squirrel can be found almost anywhere, while its nearest relative, the gray-coated round-tailed ground squirrel is largely confined to areas of drifting sand. Also, near the sand dunes where burrowing is easy, almost every bush houses a kangaroo rat, one of the most incredibly adapted creatures of the desert (Leopold, 1961, p. 83).

Sandy hills and washes have scarce surface water but enough ground water for some plant roots. It is a good habitat for small mammals and a variety of reptiles; many lizards forage in these washes. Relatively dense vegetation provides good ground cover (Creole, 1980. pp. 2-150-154).

Quite common on desert floors, on crevices and under low bushes, are the desert woodrats and the desert pocket mouse. The canyon mouse is usually found where water is plentiful, while the cactus mouse prefers sandy soils and scattered vegetation.

Several reptiles, such as the desert horned lizard, the great flat-tailed horned lizard, the desert iguana, and others, are "denizens" of sandy mesas and washes of the open desert (Jaeger, 1956. p. 79). The great flat-tailed horned lizard has a limited range and occupies dunes and other areas of windblown sand feeding mostly on ants. It is one of the endangered species and has been fully protected by the State of California since 1978 (USDI-BLM, Yuha Basin..., 1980, p. 48). The chuckwalla, the largest desert lizard, is a rock-loving lizard seldom found in the bajada areas. Another rock-loving lizard is the magic gecko. The banded rock lizard is confined mainly to the western edges of the Peninsular Ranges where it is found among the boulders and vertical walls of the gorges.

Several kinds of snakes are visible mainly in the spring, their mating season. The most common are the sidewinder, the western diamondback rattlesnake, and the common king snake.

Aside from reptiles, there are a few members of the amphibian group such as the desert tortoise, which may be found where there is adequate moisture, especially in the upland washes, springs, dunes, and oases. This species, facing extinction, is now under full protection of the California State law.

Most of the bird species are migratory, appearing only in winter, during breeding time, or migratory seasons (USDI-BLM, The California Desert..., 1980, p. 135).

Throughout the spring-flowering season, the desert and ranges are one huge food bowl for migrant as well as native birds. Along the streams one can see house finches, mockingbirds, and Le Conte's thrashers, among others. Among the rocks, running about are the canyon wrens and rock wrens. The saltbushes offer an ideal place for roadrunners to seek shelter. In and around the marshes, terns and cormorants may be seen wading.

The Gamble's quail is found not only on the desert floor but also in the canyons of upper elevations. Abert's towhees and Crissal thrashers are commonly found in the mesquite and saltbush thickets (Small, 1974, p. 234).

Of the birds of prey, the prairie falcon and the western red-tailed hawk were quite common but now are becoming rare. They inhabit the high cliffs and inter-range canyons (Jaeger, 1956, p. 102). The American kestrels are hole-nesters and inhabit elevated areas such as ocotillos, boulders, or cliffs.

A large population is that of insects. At the peak of plant growth, the desert crawls and buzzes with a large number and variety of bees, ants, wasps, beetles, and other species that appear in the spring to feed and to be fed upon (Leopold, 1961, p. 70).

The focus of life for the many animals of the desert is the water hole or "tinaja." In the evening, especially during a dry spell, water holes become crowded with nearly all-representative species.

**CHAPTER VII**

# Human Presence

Desert regions are characterized by environmental conditions that are generally adverse to human occupation. As we mentioned in the introduction, several environmental elements should be considered to understand the prehistory and history of the study area. The principal factors are water sources, lithic resources, distribution and availability of flora and fauna, and habitable regions (USDI.BLM. Yuha Basin, 1980).

Life in the desert, as elsewhere, needs water for its existence, but in the arid lands, the availability of water is unequal in time and space. The amount of water varies greatly seasonally or periodically; water may be totally lacking for considerable periods during the year only to end by torrential rains, rapidly wasted adding to soil erosion that has been taking place for centuries. Desert soils are produced almost entirely by physical weathering and are little more than fragmented rocks which contain few organic materials, and support scarcely any vegetation (Peterson, 1970, p. 15).

When one sees the desert today with its flat plains, sunbaked floors, and dry streams with scattered vegetation, it is hard to believe that at one time it had a more humid climate with many large lakes. Marine fossils dating back more than 200 million years have been found. Land fossils of 3 million years ago have been discovered in the Vallecito and Fish Creek Mountains, suggesting the existence of grasslands and wooded areas near streams where a variety of grazing animals and birds existed (Lindsay, 1973, pp. 4-7).

In many places, which are now arid bajadas and deep-set basins, running streams and bodies of water offered sanctuary to wild fauna and other small creatures. During Pleistocene time, approximately two million years ago, large animals such as the mastodon, saber-tooth tiger, and the ancient horse found this desert an agreeable place to live with plenty of food to support such populations.

Today, habitable lands are found only in certain areas, and those areas are separated by large expanses of land so unfavorable that they will remain unoccupied unless large scale development projects take place.

Man's oldest and most primitive use of the desert in his role as a food gatherer and hunter includes the harvesting of plants, which indirectly uses desert water supplies, or of desert animals which feed on those plants. The amount and distribution of flora and fauna is, therefore, one of the key elements which regulates patterns of human occupancies. For instance, when conditions of aridity are intense, trees are usually absent, but where ground water is available, as in washes and large streamways, trees often reach considerable dimensions (Jaeger, 1965, p.173). One of the best ground indicators for water is the mesquite. Where mesquite occurs in dry beds of watercourses, it has served as the surest means of successfully locating places to dig for water (Jaeger, 1965, p. 173). Roasting pits, for instance, are generally associated with desert scrub (USDI. BLM. Eastern San Diego County ..., 1980, pp. 2-56).

Weide's major premise is that the location of archeological sites can be predicted based on the relationship between site locations and a series of environmental variables within the California desert. Weide identifies certain significant assumptions that are basic to the location of sites (Weide, 1973, p. 6). First, the exploitation of natural resources was a significant determinant of land use patterns for the early native population. Second, prehistoric economic systems, although not totally understood, are related to current plant communities, physiography, geology, and hydrology. Finally, the relationship of these natural variables and the location of archeological sites is significant enough to permit prediction of site density and location within a given area.

Sources of metasedimentary, volcanic, and metavolcanic rocks were found useful in the prediction of location of sites. It is assumed that these sources could have been exploited for their lithic materials for tool making. The presence of desert pavement is also a positive indicator of site presence. Circles, trails, cairns, and lithic shops are generally found on ridges with desert pavement. Significant landforms also are good indicators. They include the base of mountains, mountain passes, valleys and ridges, granitic outcroppings, sources of water, and sheltered areas found at the base of mountains. Trails are often found on mountain passes. In the Peninsular Ranges, most of the large village sites were located within valleys or at the edge of the valleys. It is also assumed that seeps and springs are also associated with site location, as is the presence of certain useful plants. This is not surprising since it is known that mesquite, agave, pinyon, oak, and palms were exploited by the native populations as a source of food and other utilitarian purposes.

The occurrence of perennial water supply in the desert or at the desert's edge permits a more sedentary life and the development of incipient agriculture.

The date of man's appearance into Southern California is controversial because of limited archeological evidence and lack of absolute dating techniques (USDI. BLM. Yuha Basin, 1980.).

Yuha burial site skeletal remains, dated approximately 21,500 years ago, is a major finding and is the earliest man known to have existed in the area and possibly in the Western Hemisphere.

Based on the most recent archeological evidence, the earliest culture, which makes its appearance in Southern California, is the Dieguito culture. These people were hunters and gatherers. They lived about 6,000 to 10,000 years ago and the remains of their culture are the small lithic shops and isolated findings of tools in mesas and rocky terraces. This culture is associated with late Pleistocene period and the appearance of a freshwater lake.

Pinto Basin culture was probably the first true desert dwellers entering California about 5,000 to 7,000 years ago. Primarily seed gatherers, they used bedrock mortars, metates and manos.

Yuman culture appeared later, about 900 to 1000 BC, and is associated with a dry period in the Colorado Desert which forced people to move up to the higher elevations to the west or to the Colorado River to the east. During the latter part of this period, their economy was based on hunting and gathering, and developing horticulture. Evidence of this can be seen by the pottery, baskets, and other implements left behind.

Lake Cahuilla was present until approximately 500 years ago, and according to oral tradition, was full of fish, ducks, and geese. The Cahuilla, who lived in the mountains, came down to fish and hunt. As the lake subsided, they moved their villages into the valley (USDI. BLM. Cultural Resources..., 1980, p. 189). Pattern of sites are stronger for the West Mesa, located within one-quarter mile of either the relict 40-foot AMSL shoreline or Pinto Wash (USDI. BLM. Cultural Resources..., 1980, p. 177). Along the shoreline are evidences of a number of activities such as lakeside camping for the exploitation of plant resources (USDI. BLM. Cultural Resources..., 1980. p. 168). There is also evidence that seasonal habitation along Pinto Wash and Lake Cahuilla occurred during many periods (USDI. BLM. Culture Resources..., 1980. p. 172). The above-mentioned source has a good cultural sequence written by Richard L. Carrico, William Ackhardt and Jay Thasken.

At the time of European contact, the Kumeyaay occupied the area now comprising the Imperial Valley. Their economy was based on hunting, gathering, agriculture, and trade. For their agriculture, they utilized the annual flooding of the Colorado River and cultivated foodstuffs as maize, gourds, pumpkins, etc. They also used the mesquite and screwbean as a source of food (USDI. BLM. Yuha Basin, 1980). Mortars and pestles were used to make flour from dried pods and acorns. The agave was one of the most useful plants; its leaves and blooms were edible and harvested in winter. The dried fibers were used to make sandals and cordage. The yucca also had important uses; the fruit and flowers were eaten, the roots were used as soap, and the fibers to make rope and sandals (Lindsay, 1978, p. 11). See photo 29.

**Photo 29 - Yuccas, one of the most useful desert plants.**

The first Spanish entrada took place in 1774, when Juan Bautista de Anza and a small party of men arrived at Yuha Wells. The following year, de Anza again visited the area, this time with a group of colonists and a herd of cattle. The Yuma Trail passes east-west through the northern part of the Yuha Basin. Another road was constructed later between the original route and Yuha Wells (USDI. BLM. Yuha Basin, 1980).

But Spaniards viewed the California desert as little more than a forbidding obstacle over which to travel between northern Mexico and Arizona to California. The trail became known as the Southern Emigrant Trail, the Sonora Road, or the Colorado Road (USDI. BLM. Cultural Resources..., 1980. p. 40).

With the transfer of lands to the United States in the mid-19th century, the use of the land intensified. After initial forays by Mormon settlers, miners and soldiers, a growing stream of emigrants bound for coastal California came through the desert, protected by military outposts along the route (USDI. BLM. Plan Alternative..., 1980, p. 1).

International boundary survey party member, Dr. William P. Blake, was responsible for naming the "barren and desolate lands" the Colorado Desert (USDI. BLM. Cultural Resources..., 1980, p. 40). This document has an excellent historical sequence put together by Frank Norris and Terri Jacques.

Railroad facilities, mining operations, and livestock grazing occurred over a wide expanse of the desert. At the turn of the century, the construction of a canal from the Colorado River transformed the central portion of the desert into one of the most agriculturally productive areas of the country. The All-American Canal was constructed under public works criterion in the 1930s. Military concerns took priority during WWII, and the desert experienced one more impact as General Patton's troops trained across vast expanses of the area, preparing for a North African campaign. Later, the Army and the Navy reserved large portions of the land for training and testing of military weapons. By the middle of the 20th century, major stretches of the desert were permanently entrenched: livestock, agriculture, mining, military operations, major transportation arteries and the growth of permanent settlements (USDI. BLM. Plan Alternative..., 1980, p. 1).

The major transportation routes in the region include three state highways and two county highways which transverse the region from east to west. Interstate 8 and State Highway 98 cross the Yuha Basin, 1.5 miles and 8.5 miles south of Plaster City respectively, joining at Ocotillo to the west. Entering the Ocotillo area from the northwest and the Anza-Borrego Park is County Highway S2, the historical Imperial Highway, which continues east of Ocotillo as County S80, the main highway through Coyote Wells, Plaster City and Dixieland; three small communities in the region. State Highway 86 borders the area of study to the east, departing from Brawley and moving in a north-west direction at the west edge of the Salton Sea toward Coachella and Indio.

At the junction of State Highway 86 and San Felipe Creek is the beginning of State Highway 78 which crosses Ocotillo Badlands, Ocotillo Wells, Lower Borrego Valley, and Sentenac Canyon before going on to Julian and the Southern California Coast. It has been a major corridor of travel since the late 1800s (Lindsay, 1978, p. 34). County road S22, the most northern in the study area, connects Salton City with Borrego Springs, and it is also known as Borrego-Salton Seaway. A series of unmaintained dirt and gravel roads crisscross the undeveloped open desert areas between these highways. Their number has increased since the early 1960s, the beginning of the dune buggy culture.

The San Diego and Arizona Eastern Railroad Company, a subsidiary of the Southern Pacific Railroad Co., operated a freight line through the region which connected San Diego with El Centro, via Tijuana, Mexico. The railroad ran through Plaster City and served the U.S. Gypsum Co. plant. In September 1976, Hurricane Kathleen destroyed Interstate 8 and the trail line. Since then, the highway was reconstructed but not the rail line. Plans were to open it again in 1981. Work on the California side is almost completed, yet still needed at the bridges in Baja California in the vicinity of Tijuana. The old Overland Stage Route, a desert road from Plaster City northwest into Anza-Borrego State Park, has historical value. Along S2, in the vicinity of Agua Caliente Springs, there are still remnants of the Great Southern Route of the 1860s. See photo 30.

**Photo 30 - The Butterfield State Route sign on County Highway S2.**

A 25-mile narrow gauge railroad line still runs across the West Mesa from Plaster City to the northwest where Gypsum beds are at the base of the Fish Creek Mountains. See photo 31.

**Photo 31 - Railroad line from gypsum mine to Plaster City.**

The agricultural development of the Imperial Valley for the last 80 years has had a relatively insignificant impact on the area under investigation. Communication corridors and transmission lines were established across the area from the irrigated lands pointing east to the mountains (where women and children were sent to spend summers), and eventually to the coast. The newest of these lines will be an electric transmission line from the Colorado River to San Diego.

U.S. Gypsum Co. created Plaster City close to the gypsum beds. It is a relatively small operation but represents the largest portion of the non-agricultural economic activity of Imperial County. A cement plant, TXI, has been approved for construction in the near future.

The largest portion of the population in our study area appears to be people who have retired from overcrowded, noisy, and polluted urban centers who have learned to live with the heat, winds, and limited water resources. These residents appear to have a good relationship with "snowbirds", northerners who come into the warmer area to spend the winter months. There is some conflict, however, between these property owners and the thousands of "weekend-holiday" urban dwellers who enter the area to use their recreational vehicles. This yearly conflict begins in September during the Labor Day weekend, and ends in May around Memorial Day weekend.

In the middle of the conflict is the Imperial County government and the Bureau of Land Management since they are both responsible for the present and future land use characteristics of the area. The county governments are responsible for deciding whether new areas will open for residences in relation to availability of ground water resources. It is also their responsibility to decide on future economic development in the area, also weighing the water resources available.

Over the past five years, the Bureau of Land Management has been working on a desert land use plan for public lands. Public participation and input from special local and state interest groups has occurred, and in the process the plan is still changing. There is now, however, some orderly use of public lands.

The total permanent population is scattered in a series of communities which are slowly growing, especially in mobile home residencies. Any future growth will largely depend on county government decisions and the decreasing quality of the Salton Sea (L. Fabian, personal communication, Imperial County Planning Department, 1985).

The area under study within Imperial County is called West Imperial County Census Division, and in the Special Census of 1975 had a population of 1340. The unofficial 1980 count of 1882 is probably low but represents a 40% increase.*

| AREA | 1975 | 1980 |
| --- | --- | --- |
| West Imperial County Census Division | 1340 | 1882 |
| West Shores | 887 | ----- |
| Salton City | ----- | 775 |
| Desert Shores | ----- | 446 |
| Ocotillo | 223 | ----- |
| Yuha Desert | 91 | ----- |
| Plaster City | 96 | ----- |

This permanent population is augmented with desert and Salton Sea recreation users. For the last two years the numbers have declined (R. Pollock, Imperial County Parks and Recreation, personal communication, 1985).

Records kept by this office over the last seven years, based on weekend and holiday flights over the area and land enumeration, show relatively heavy recreational use of the area.

| AREA | WEEKEND COUNTS |
| --- | --- |
| Salton City and Desert Water Shore Front- Summer | 3000 to 5000 |
| Highway 78 - overflow from San Diego County- Fall, Winter, Spring | 2000 |
| Parachute Area - Fall, Winter, Spring | Less than 100 Interstate 8 |
| Plaster City, Ocotillo Painted Gorge and Mountain Top – Fall, Winter, Spring | 2000 to 3000 |
| Race Areas - during races | 2000 |
| Yuha Basin – Fall, Winter, Spring | Less than 500 |

The weekend recreational population far outnumbers the permanent population, but their numbers have been declining and will continue to decrease due to higher energy costs and the decreasing quality of the Salton Sea. At the present, unless some serious steps are taken, the biological death of the Salton Sea appears inevitable (L. Fabian, Imperial County Planning Department, personal communication, 1985).

The danger of destruction of anthropological evidence by other land uses – recreation, off-road vehicles, and agriculture – requires that intensive fieldwork, analysis, and documentation take place before it is too late.

The total permanent population will continue to increase as more people in urbanized areas find it more challenging to live in the cities, both environmentally and economically. The seasonal and weekend temporary population will continue to invade the area, largely because recreational opportunities are more affordable and less restricted than in the coastal areas.

Despite somewhat limited water resources, capital investors will continue to come into the area because of open space, good transportation systems, proximity to markets on the coast, less restricted land use and zoning regulations.

Under these present and future land use pressures, it will be up to the county government and federal agencies, with the help of concerned citizens, to safeguard the fragile ecological balance of these desert areas. Consideration of the future of this area is critical. Careful and serious planning is needed.

* The census enumeration defines residency as six months and one day, therefore many residences were reported as unoccupied. In Desert Shores, 517 dwellings (mostly trailers) were recorded in 1980 and of these, 285 (55%) were reported as unoccupied (L. Fabian, Imperial County Planning Department, personal communication, 1985).

# APPENDICES

APPENDIX  A

Desert Habitats

| Habitat | Elev. Range | Plants | Animals |
|---|---|---|---|
| DESERT WASH - found in dry water courses leading from canyons of desert mountains and in badlands washes. (These washes support a greater diversity of plants and animals than does the surrounding desert because of the concentration of water received and carried at certain times of the year). | 3000 to below sea level | Smoke tree, ironwood, palo verde, desert willow, cheesebush, creosote, burroweed, brittle-bush, mesquite, mistletoe, indigo bush, bunch grass, desert lavander, loco weed, trumpet flower, desert aster and various annual flowers including bladderpod, desert lily, verbena, dune and desert primrose, spectacle pod and sand lupine. | Jackrabbit, cottontail, desert wood rat, coyote, badger, kit fox, gray fox, antelope ground squirrel, phainopepla, verdin, zebra-tailed lizard, desert iguana and sidewinder. These animals are found in and adjacent to desert washes. |
| LOW DESERT AND DESERT SLOPE- found in desert valleys and on slopes. Topography is rough, steep and broken on the mountain slopes. Deep canyons and steep ridges are the rule. The slopes end rather abruptly at the desert floor. | 3000 to sea level | Plants are generally sparse, but grow over wide areas. Common plants include creosote, burroweed, indigo bush, brittle-bush, ocotillo, cholla, barrel cactus and desert thorn. | Antelope ground squirrel, jackrabbit, white-throated woodrat, Merriam's kangaroo rat, little pocket mouse, road runner, cactus wren, raven, Say's phoebe, zebra-tailed lizard, desert iguana, sidewinder, red racer and, in dune areas, kit fox and desert kangaroo rat. On mountain slopes bighorn sheep, bobcats and a few deer are found. |
| CHAPARRAL - found on higher desert slopes. A dense cover of shrubs, some reaching 15 feet in height, is characteristic. Water runoff courses often have thick stands of scrub oak. | 3000- 5500 | Chamise, scrub oak, ceanothus, manzanitas, sumac and yuccas. On desert slopes these will grade into desert species such as brittle-bush and various cacti. | Mule deer, coyote, gray fox, bobcat, rabbit, mountain quail, scrub jay, poor-will, western fence lizard, western rattlesnake and numerous invertebrates. |

OVER

CONTINUE

| Habitat | Elev. Range | Plants | Animals |
| --- | --- | --- | --- |
| PINYON JUNIPER- found on desert slopes of the Santa Rosa, Laguna, In-Ko-Pah and Jacumba mountains and other localities. Consists of scattered trees 10-30 feet tall in open stands, mixed with shrubs and grading into a desert slope chaparral. Topography is gentle to very steep, with rocky areas common. | below 6000 | Pinyon, juniper, scrub oak, sumac, cholla, California mountain joint fir, mountain mahogany, manzanita, Parry nolina, desert agave and buckwheat. | Coyote, jackrabbit, Beechey ground squirrel, piñon mouse, desert woodrat, California thrasher, poor-will, black-throated gray warbler, canyon and rock wrens, red-tailed hawk, king snake, gopher snake. |

From: Lindsay, L. *The Anza-Borrego Desert Region.* Wilderness Press, Berkeley, CA., 1978, p. 9.

APPENDIX B

TYPICAL FLORA AND FAUNA OF AREA UNDER STUDY*

| Common Name | Generic Name |
| --- | --- |
| FLORA | |
| Arizona Catclaw | acacia gregii |
| Arrowweed | pluchea sericea |
| Barrel Cactus or Bisnaga | ferocactus wislizeni |
| Beavertail Cactus | opuntia basilaris |
| Black Walnut | juglans californica |
| Bladderpod | isomeris arborea |
| Buckhorn Cholla | opuntia acanthocarpa |
| Bursage or Burroweed | franseria dumosa |
| California Poppy | eschscholzia parishii |
| Chamise | adenostoma fasciculatum |
| Cheese Bush | hymenodea salsola |
| Chollas | opuntia sp. |
| Chuparosa | beloperone californica |
| Cinch Weed | pectis paposa |
| Common Reed | phragnites communis |
| Cottontops | echinocactus polycephalus |
| Cottonwood | populus fremonti |
| Coyote Gourd | cucurbita palmata |
| Creosote Bush | larrea divaricata |
| Crucifixion Thorn | castela emoryi |
| Desert Agave | agave deserti |
| Desert Encelia or Brittlebush | encelia farinosa |
| Desert Holly | atriplex hymenelytra |
| Desert Primrose | oenothera deltoides |
| Desert Willow | chilopsis linearis |
| Elephant Tree | bursera microphylla |
| Fairy Dusters | calliandra eriophylla |
| Four O'Clocks | mirabilis froebelii |
| Four-Winged Saltbush | atriplex canescens |
| Galleta Grass | hilaria rigida |
| Honey Mesquite | prospopis glandulosa |
| Hedgehog Cactus | echinocereus engelmannii |
| Indian Wheat | plantago purshii |
| Ironwood | olneya tesota |
| Jimson Weed | datura meteliodes |
| Jumping Cholla or Teddy Bear Cholla | opuntia begelovii |
| Juniper Pine | junipero californica |
| Live Oak | quercus wislisenii |
| Lupine | lupinus excubitus |
| Manzanita | arctostephylos pungens |
| Mohave Yucca | yucca schidigera |
| Mormon Tea | ephedra californica |
| Ocotillo | fouquieria splendens |

*These are not intended to be exhaustive lists of all animals and plants
in the area. Only some of the most common species are listed here.

| Palo Verde | cercidium floridum |
| Parry Nolina | nolina parryae |
| Phacelia | phacelia crenulata |
| Prickly Pear | opuntia sp. |
| Purple Sage | salvia leucophylla |
| Rock Nettle | eucnide rupestris |
| Rush Milkweed | asclepias sublata |
| Screwbean Mesquite | prosopis pubescens |
| Scrub Oak | quercus dumosa |
| Silver Cholla | opuntia echinocarpa |
| Smoke Tree | dalea spinosa |
| Snakeweed | gutierretzia sarathrae |
| Tall Cat-tail | typha latifolia |
| Tamarisk | tamarix sp. |
| Verbena or Sand Verbena | abronia villosa |
| Washingtonia Palm | washingtonia filifera |
| Wild or Imperial Buckwheat | eriogonum fasciculatum |
| White Sage | salvia apiana |
| Yucca | yucca bacata |

FAUNA

Mammals

| Antilope Ground Squirrel | ammos permophilus leucurus |
| Audubon Cottontail | sylvilagus auduboni |
| Badger | taxidea taxus |
| Black-tailed Jackrabbit | lepus californicus |
| Bobcat | felis rufus |
| Brush Mouse | peromyscus boylei |
| Cactus Mouse | peromyscus eremiscus |
| Canyon Mouse | peromyscus crinitus |
| Coyote | canis latrosn estor |
| Desert Cottontail | sylvilagus bachmani |
| Desert Kit Fox | vulpes macrotes |
| Desert Pocket Mouse | perognathus penicillatus |
| Desert Woodrat | neotoma lepida |
| Kangaroo Rats | dipodomis agilis, d. merriami. d. deserti |
| Mountain Lion | felis concolor |
| Mule Deer | odocoileus hemionus |
| Peninsular Bighorn Sheep | ovis canadensis cremnobates |
| Racoon | procyon lotor |
| Ringtail | bassariscus astutus |
| Round-tailed Ground Squirrel | citellus terreticaudos |
| Spring Pocket Mouse | perognathus spinatus |
| Striped Skunk | mephitis mephitis |

Amphibians and Reptiles

| Common Kingsnake | lampropeltis getulus |
| Colorado Desert Sidewinder | crotalus cerastes |
| Desert Blackheaded Snake | tantilla planiceps transmontana |
| Gopher Snake | pitouphis melanoleucus annectens |

Red-diamond Rattlesnake                     crotalus ruber
Red Racer                                   masticophis flagellum piceus
Spotted Leaf-nosed Snake                    phyllorhynchus decurtatus
Western Blind Snake                         leptotyphlops humilis
Western Diamondback Rattlesnake             crotalus atrox

Banded Rock Lizard                          streptosaurus mearnsi
Chuckwalla                                  sauramalus obesus
Colorado Fringe-toed Lizard                 uma notata
Desert Tortoise                             gopherus agassizi
Desert Iguana                               dipsosaurus dorsalis
Desert Horned Lizard                        phrynosoma m. callii
Desert Banded Gecko                         coleonyx variegatus
Flat-tailed Horned Lizard                   phrynosoma m'callii
Great Basin Whiptail Lizard                 cnemidophorus tigris
Leopard Lizard                              gambilia wislizanii
Magic Gecko                                 anarbylus switaki
Red-spotted Toad                            bufo punctatus
Rock Lizard                                 petrosaurus mearshi
Slender Salamander                          fatrachoseps attenuatus

Birds
Abert's Towhee                              pipilo aberti
American Kestrel                            falco sparverius
Audubon's Warbler                           dendroica auduboni
Barn Owl                                    tyto alba
Burrowing Owl                               speotyto cunicularia
Bendire's Thrasher                          toxostoma bendirei
Black-tailed Gnatcatcher                    polioptila melanura
Cactus Wren                                 campylorhynchus brunneicapillus
California Quail                            lophortyx californicus
California Yellow-billed Cuckoo             coccyzuz americanus
Canyon Wren                                 catherpes mexicanus
Common Flicker                              colaptes cafer
Common Raven                                corvus corax
Costa's Hummingbird                         calypte costae
Crissal Thrasher                            toxostoma dorsale
Desert Song Sparrow                         melospiza melodia
Double-crested Cormorant                    phalacrocorax auritus
Gamble's Quail                              lophortyx gambellii
Golden Eagle                                aquila chrysaetos
Great Horned Owl                            bubo virginianus
House Finch                                 carpodacus mexicanus
Le Conte's Thrasher                         toxostoma lecontei
March Hawk                                  circus cyaneus
Mockingbird                                 mimus polyglottus
Mourning Dove                               zenaida macroura
Phainopepla                                 phainopepla nitens
Red-tailed Hawk                             buteo jamaicensis
Roadrunner                                  geococcy californianus
Rock Wren                                   salpinctes obsoletus

| Vermillion Flycatcher | pyrocephalus rubinus |
| Western Kingbird | tyrannus verticalis |
| White-winged Dove | zenaida asiatica |
| Yellow Warbler | dendroica petechia |
| Wilson's Warbler | wilsonia pasilla |

# GLOSSARY

Acidic:

refers to the ph scale; substances that test below 7 ph. (See also <u>basic</u>).

Abiotic:

non-living material which comprises the physical environment of the biotic elements.

Alkali:

soluble mineral salts present in soils, mostly in arid lands.

Alluvium:

sediments deposited by rushing water, e.g.: sediments from washes.

Alluvial fan:

alluvium found at the base of mountains. Gets its name from its fan-like structure or pattern.

Anticline:

upfolded rock, a result of diastrophism.

Arroyo:

small watercourse or gulch usually dry except after heavy rains.

Badlands:

landforms characterized by little or no vegetation, high erosion, and highly dissected ravines.

Bajada:

coalescing rows of alluvial fans.

Barchans:

crescent-shaped sand dunes with convex sand windward and concave side leeward.

Basic:

refers to the ph scale, substances that test above 7 ph. (See also <u>acidic</u>).

Basal:

of, at, or forming the base.

Biome:

a biogeographical or ecological subunit comprised of a number of communities.

Biotic:

that which is living.

Bolsones:

a closed, mountain-surrounded basin having a playa for drainage.

Breccia:

rock composed of angular fragments of older rocks melded together.

Chimneys:

wind eroded sandstone deposits infiltrated with calcium carbonate; landforms referred to as such by their shape.

Cienega:

a swamp or marsh, especially one formed and fed by springs.

Caliche:

end result of an ongoing moisture/evaporation process with

improper drainage; a thick, cement-like crusting of carbonates and salts.

Concretions:  a rounded mass of mineral matter occurring in sandstone, clay, etc., often in concentric layers about a nucleus.

Conglomerates:  sedimentary rocks composed of rounded and waterworn pebbles or the like embedded in a finer cementing material.

Crenelate:  notched and/or groved.

Crown:  first vegetation to sprout after a fire; usually plants with extensive root systems, e.g.: perennials.

Culminate:  the highest point of an intermont valley or canyon found at the base of a mountain.

Comminute:  to pulverize or triturate.

Declivity:  a downward slope, as of ground.

Deflation:  the first step of wind erosion; the lifting and removal of sand and soil material.

Dendritic:  drainage pattern where many tributaries are present giving the appearance of a veiny leaf.

Denudation:  the stripping of all flora and fauna from the landscape.

Desert Varnish:  the dark, lustrous coating or crust, usually of manganese and iron oxides, that form on rocks, pebbles, etc, in the desert.

Detrital:  composed of detritus, desintegrated material, debris.

Diastrophism:  the action of the forces which cause the earth's crust to be deformed, producing continents, mountains, changes in levels, etc.

Dikes:  a long, narrow, cross-cutting mass of igneous or eruptive rock intruded into a fissure in older rocks.

Dreikanter:  a pebble or boulder having three faces formed by the action of windblown sand.

Ephemerals:  short-lived plants, e.g.: annuals.

Evanescent:  tending to become imperceptible.

Exfoliation:  a process of erosion tending to split or swell into scaly aggregates, end result of extreme temperatures.

Exudation:       the breaking down and carrying away by water through porous
                 materials.

Fault:           the displacement of a section of rock in relation to
                 another.

Ferruginous:     iron-bearing: of the color of iron rust.

Foliated rock:   rock split into thin sheets, often a result of exfoliation.

Geomorphology:   the study of characteristics and developments of landforms.

Gneiss:          banded and foliated rock; exfoliated metamorphic, some-
                 times rich in feldspar and quartz, others rich in horn-
                 blende or mica.

Gradient:        a ratio between vertical rise and horizontal distance.

Graben:          a portion of the earth's crust, bounded on at least two
                 sides by faults, that has been moved downward in relation
                 to adjacent portions.

Halophytes:      salt-resistant plants.

Humic:           the conditions of humus soils, having high nutrients,
                 somewhat acidic, and having a high count of microbiotic
                 materials.

Hydration:       to combine chemically with water.

Hydrophytes:     plants suited to high water conditions.


Igneous rocks:   highly dense rocks with relative even texture, often crys-
                 taline, produced under intense heat, generally of volcanic
                 origin.

Inundate:        to flood.

Lacustrine:      of or relating to lakes; formed by lakes.

Leeward:         the side of a mountain, hill or dune sheltered from the
                 wind.

Lichen:          any group of lichens that is composed of fungus in symbio-
                 tic relationship with an algae.

Malpais:         "Bad Country", refers to land recently covered with lava
                 and devoid of vegetation.

Mesophytes:      plants suited to moderate water conditions.

Metamorphic:       an alteration in the physical structure of igneous rocks
                   and/ or sedimentary rocks due to natural agencies such as
                   heat or pressure.

Orogenic:          relating to the process of mountain making or upheaval.

Outcrop:           a rash-like appearance on a typical landscape; a salient
                   feature.

Patina:            the film or coloring on rock surfaces, the end result of
                   an oxidation process involving water and soils having
                   ferric oxide.

Pediments:         a recessed  mountain  slope.

Perennials:        plants having a life cycle of more than two years.

Ph:                a coefficient indicating acidity/alkalinity ph 7 is
                   neutral; less than 7, acid; more than 7, basic.

Playa:             a low basin area where silt and clay are carried;
                   site of a seasonal or temporal lake.

Population:        a biogeographical or ecological unit; a number of like
                   individuals within a community.

Porous:            full of pores; permeable by water, air, etc.


Primary succession: the first and often dominant specie to appear after
                   denudation.

Reticulates:       divided with lines, a network.

Rift:              the surface indication of a fault line.

Rills:             narrow, eroded groves of topsoil, from 1" to 2" deep.

Riparian:          situated in the bank of a river or other body of water.

Salient:           an outcrop or otherwise nonconformed feature in the land-
                   scape.

Saltation:         leaping movement of sand or gravel by the action of wind
                   or running water.

Scarp:             the leading edge of a fault line.

Schist:            crystalline rocks whose constituent minerals have more or
                   less parallel or foliated  arrangement, due to metamorphic
                   action.

Sedimentary:       type of rock formed by the deposition of sediments, less
                   dense than igneous and with a pervasive cement like element,
                   e.g.: calcite, quartz, ferric oxide, etc.

Siliceous:         silica deposits; type of landform.

Silt:              earthy matter carried by running water and deposited
                   as sediments. Clastic particles 1/16 mm to 1/256 mm
                   in diameter.

Sink:              the lowest portion of a basin or graben which receives
                   drainage.

Slope:             the degree of vertical rise.

Subsidence:        to sink to a lower level.

Subsoil:           the stratum of earthly material immediately under the
                   surface soil.  For this study about 10" to 40" below the
                   surface.

Substratum:        (for this study) soils found below 40"; parent material
                   which resembles the unaffected materials of which subsoils
                   and surface soils were made.

Surface soils:     (for this study) soils found between 0" and 10" in depth.

Talus:             an accumulation of rock fragments which are found at
                   the base of a rock face.

Tessellate:        stones found to be arranged in mosaic-like patterns.

Tinaja:            water hole.

Travertine:        a form of limestone deposited by springs.

Trough:            any long depression between two ridges.

Tufa:              a porous limestone formed from calcium carbonate deposited
                   by springs.

Warp:              an uplift or sunken portion of the earth crust; nonfaulted
                   uplift.

Windrows:          debris gathered by the wind.

Windward:          that side of a mountain, hill or dune which faces the wind.

Winnow:            to separate lighter particles by the action of the wind.

Xerophyte:         plants suited to drought conditions.

# BIBLIOGRAPHY


Aschmann, Homer
    1967        "Purpose in the Southern Landscape". The Journal of Geography, Vol. 66, September 1967, pp. 311-317.

Atwood, Wallace
    1940        The Physiographic Provinces of North America. University of California Press, Berkeley, CA.

Bailey, Harry P.
    1966        The Climate of Southern California. University of California Press, Berkeley, CA.

    1954        "Climate, Vegetation and Land Use in Southern California". Geology of Southern California. State of California, Department of Natural Resources, Division of Mines, Bulletin 170, Vol. 1.

Bakker, Elna
    1972        An Island Called California. University of California Press, Berkeley, CA.

Balls, Edward K.
    1962        Early Uses of California Plants. University of California Press, Berkeley, CA.

Bateman, Paul C. and William P. Irwin
    1954        "Tungsten in Southern California". Geology of Southern California. State of California Department of Natural Resources, Division of Mines, Bulletin 170, Vol. 1.

Baugh, Ruth E.
    1955        The Geographic Regions of California: A Syllabus for the Study of the Geography. Pacific Books, Palo Alto, CA.

Bean, Lowell and Katherine Saubel
    1961        "Cahuilla Ethnobotanical Note: The Aboriginal Use of Oak". Archaeological Survey Annual Report, Department of Anthropology, University of California, Los Angeles.

Bennet, C.A.
    1975        Climate of the Southern Desert Air Basins. California Air Resources Board, Sacramento, CA.

Benson, L.
    1979        The Native Cacti of California. Stanford Press, Stanford University, CA.

Black, William E.
    1974        A Geophysical Investigation of the Yuha Desert, Imperial County. Master Thesis. University of California, Riverside.

Blackwelder, Eliot
    1954        "Geomorphic Processes in the Desert". _Geology of
                Southern California_. State of California, Department
                of Natural Resources, Division of Mines, Bulletin 170,
                Vol. 1.

Bowers, Stephen
    1901        _Reconnaissance of the Colorado Desert Mining District_.
                California State Mining Bureau, San Francisco, CA.

Burns, Helen
    1952        _The Salton Sea Story_. Desert Magazine Press, Palm Desert,
                CA.

Chaetham, Norden and Robert Haller
    N.D.        _An Annotated List of California Habitat Types_. University
                of California Land and Water Reserve System. Unpublished.

Childers, Morris W.
    N. D.       "Sandstone Chimmneys in the Imperial Formation".
                Reprint. Imperial Valley, CA.

Cline, Lora L.
    1979        _The Kwaaymii: Reflections on a Lost Culture_. IVC Museum
                Society, El Centro, CA.

Cornett, Jim
    1975        _Wildlife of the Southwestern Desert_. Nature Trails,
                Desert Hot Springs, CA.

Creole Corporation
    1980        _Final Environmental Impact, Portland Cement Plant_.
                Imperial County, CA.

Dibble, W.T.
    1954        "Geology of the Imperial Valley Region, California".
                _Geology of Southern California_. State of California,
                Department of Natural Resources, Division of Mines,
                Bulletin 170, Vol. 1.

Dunbier, Roger
    1968        _The Sonoran Desert, Its Geography, Economy and People_.
                University of Arizona Press, Tucson, AZ.

Durrenberger, Robert
    1968        _Elements of California Geography_. National Press Books,
                Palo Alto, CA.

    1959        _The Geography of California in Essays and Readings_.
                Breusters Publishers, Los Angeles, CA.

1972        Patterns on the Land: Geographical, Historical, and
            Political Maps of California. National Press Books,
            Palo Alto, CA.

Fabian, Leonard
     1980   Imperial County Planning Department. Personal communication.

Federal Writers Project of the Work Program Administration for the
State of California
     1972   California- A Guide to the Golden State. Hastings House
            Publishers, New York.

Gray, Margaret
     1973   Glossary of Geology. American Geological Institute,
            Washington, D.C.

Griffin, Paul and Robert Young
     1957   California - The New Empire State. Fearson Publishers,
            San Francisco, CA.

Gudde, Erwin
     1952   California Place Names. University of California Press,
            Berkeley, CA.

Fischer, P.
     1973   "Imperial Valley Geology". APPG Guide Book,  Trip No. 2.

Hartman, David N.
     1968   California and Man. Wm. C. Brown Co. Publishers, Dubuque,
            Iowa.

Heizer, Robert Fleming
     1971   The California Indians. University of California Press,
            Berkeley, CA.

Hendricks, William
     1971   "Developing San Diego's Desert Empire". The Journal of
            San Diego History. San Diego, CA. Summer 1971.

Hicks, Sam
     1966   Desert Plants and People. Naylor Co., Dallas, Texas.

Hill, D.P., P. Mowinckel and L.G. Peake
     1966   "Earthquake, Active Faults, Geothermal Areas in the
            Imperial Valley, CA.". Science, Vol. 188, p. 1306.

Hinds, Norman E.
     1952   Evolution of the California Landscape. State of California,
            Department of Natural Resources, Division of Mines,
            Bulletin 158.

Hodge, Carle and Peter Dwisberg
      1965          Aridity and Man: The Challenge of the Arid Lands in
                    the United States . AAAS, Washington, D.C., Publication
                    74.

Imperial County Planning Department
      1975          Population. Bulletin. Imperial County, CA.

Imperial Irrigation District
      1967          Report for General Soil Map. Imperial County, CA.

      1974          Water Report. Unpublished. Imperial County, CA.

Jaeger, Edmund C.
      1965          The California Deserts. Stanford University Press,
                    Stanford, CA.

      1957          The Denizens of the Desert.  Stanford University Press,
                    Stanford, CA.

      1961          Desert Wildlife. Stanford University Press, Stanford,
                    CA.

      1957          The North American Desert. Stanford University Press,
                    Stanford, CA.

Jahns, Richard
      1954          "Geology of the Peninsular Ranges, California".
                    Geology of Southern California. State of California,
                    Department of Natural Resources, Division of Mines,
                    Bulletin 170, Vol. 1.

Jensen, William and Frank Salisbury
      1972          Botany: An Ecological Approach. Wadworth Publishing Co.,
                    Belmont, CA.

Lantis, David, Rodney Steiner, and Arthur Karinen
      1977          California: Land of Contrasts. Kendall-Hunt Publishing
                    Co., Dubuque, Iowa.

Lawrence Livermore Laboratory
      1976          Imperial Valley Environmental Project - A Description
                    of the Imperial Valley, California for the Assesment of
                    Impacts of Geothermal Energy Development. University of
                    California, Livermore, CA.

Lee, William
      1963          The Great California Deserts. Putnam Press, New York.

Leopold, A. Starker
      1961          The Desert.  Life Nature Library, New York.

Lindsay, Diana Elaine
        1973        Our Historic Desert. Union-Tribune Publishing
                    Co., San Diego, CA.

Lindsay, Lowell and Diana
        1978        The Anza-Borrego Desert Region. Wilderness Press,
                    Berkeley, CA.

Loeltz, O.J., B. Ireland, J.A. Robinson, and F.H. Olmsted
        1975        Geohydrologic Reconnaissance of the Imperial
                    Valley, California. Geological Survey. Professional
                    Paper 486, United States Government Printing Office,
                    Washington, D.C.

Miller, William J.
        1957        California Through the Ages: A Geologic Story of a
                    Great State. Westernlore Press, Los Angeles, CA.

Morton, P. R.
        1977    .   "Geology and Metal Resources of Imperial County, CA".
                    California Division of Mines and Geology. County
                    Report 7.

Munz, Philip A.
        1962        California Desert Wildflowers. University of California
                    Press, Berkeley, CA.

Munz, Philip A. and David Keck
        1970        A California Flora.  University of California Press,
                    Berkeley, CA.

Murphy, Edith Van Allen
        1959        Indian Uses of Native Plants. Mendocino County
                    Historical Society, CA.

Parker, Horace
        1963        Anza Borrego Desert Guide Book. Paisano Press,
                    Balboa, CA.

Pepper, Choral
        1972        Guidebook to the Colorado Desert of California.
                    Ward Ritchie Press, Los Angeles, CA.

Pollock, Richard    Director, Imperial County Parks and Recreation Depart-
        1980        ment. Interview.

Pourade, Richard
        1966        History of San Diego. Union-Tribune Publishing
                    Co., San Diego, CA.

Reed, Ralph D.
    1933        Geology of California. American Association of
                Petroleum Engineers, Tulsa, Oklahoma.

Reinman, Fred, D. L. True, and Claude Warren
    1960        Archaeological Remains from Rock Shelters Near
                Coyote Mountain, Imperial California. Annual
                Report, Archaelogical Survey. Department of
                Anthropology-Sociology, University of California,
                Los Angeles.

Rogers, Malcom
    1968        "An Outline of Yuman Prehistory". Southwestern
                Journal of Anthropology. Vol. 1, No. 2, pp.
                167-198.

    1966        Ancient Hunters of the Far West. Union-Tribune
                Publishing Co., San Diego, CA.

Salitore, Edward
    1971        California - Past, Present, and Future. California
                Almanac Co., Lakewood, CA.

Sharp, Robert P.
    1972        Field Guide to Southern California. Wm. C. Brown
                Co., Dubuque, Iowa.

Shreve, Forrest
    1951        Vegetation of the Sonoran Desert. Carnegie Institute
                Publications, Washington, D.C.

Small, Arnold
    1974        The Birds of California. Winchester Press, New
                York.

Stanley, G.M.
    1962        "Prehistoric Lakes in the Salton Basin". Geological
                Society of America Special Paper 73. (Abstract for
                1962) pp. 249-250.

State of California
    1979        Population Estimates for California Counties. Popula-
                tion Research Unit, Sacramento, CA.

Stebbins, Robert
    1972        Amphibians and Reptiles of California. University
                of California Press, Berkeley, CA.

Sutton, Ann and Myron Sutton
    1966        The Life of the Desert. McGraw-Hill Book Co., New
                York.

Thomas, Robert
    1963            "The Late Pleistocene 150-foot Freshwater Beachline
                    of the Salton Sea Area". Southern California Academy
                    of Science Bulletin, Vol. 62, No. 1, pp. 9-17.

Thomas, William L.
    1961            "Competition for a Desert Lake: The Salton Sea,
                    California". The California Geographer, Vol. 2,
                    pp. 31-40.

Thornbury, William
    1958            Principles of Geomorphology. John Wiley and Sons,
                    New York.

    1965            Regional Geomorphology of the United States. John
                    Wiley and Sons, New York.

USDA. Soil Conservation Service and U.S. Forest Service
    1973            Soil Survey of San Diego County. San Diego, CA.

USDA. Soil Conservation Service, University of California Agricultural
Experimental Station, and Imperial Valley Irrigation District
    N.D.            Imperial Valley Soil Survey Report. Unpublished Draft,
                    1962-1975.

USDI. Bureau of Land Management
    1980            The Archaelogy and History of the McCain Valley
                    Study Area. Eastern San Diego County, CA.: A Class
                    II Cultural Resources Inventory. Riverside, CA.

    1980            The California Desert Conservation Area. Plan Alterna-
                    tive and Environmental Impact Statement. California
                    State Office, Sacramento, CA.

    1980            California Desert Conservation Area. Final Environ-
                    mental Impact Statement and Proposed Plan.
                    California State Office, Sacramento, CA.

    1980            Cultural Resources Inventory of the East Mesa and
                    West Mesa Regions of Imperial Valley, CA.
                    Riverside, CA.

    1980            Proposed Livestock Grazing and Wilderness Management
                    for the Eastern San Diego County Planning Unit.
                    Environmental Impact Statement, Riverside, CA.

    1980            Yuha Basin Proposed Geothermal Leasing. Final
                    Environmental Assessment Record. Riverside District
                    Office, Riverside, CA.

USDI. Geological Survey
    1968            The Borrego Mountain Earthquake of April 9, 1968.

United States Government Printing Office,
Washington, D.C.

Van Devender, Thomas and Goeffrey Spaulding
    1979       "Development of Vegetation and Climate of the
Southwestern United States". Science, Vol. 204,
No. 18, pp. 701-710, May 1979.

Van De Kamp, P.C.
    1973       "Holocene Continental Sedimentation in the Salton
Basin, California - A Reconnaissance". Geol. Soc.
Am. Bull., Vol. 84, No. 3, p. 827.

Walton, Kenneth
    1969       The Arid Zones. Aldine Publishing Co., Chicago, Ill.

Watkins, T. H.
    1973       The Grand Colorado, the Story of a River and Its
Canyons. American West Publishing Co., Palo Alto, CA.

Weide, Margaret L.
    1973       "Archaeological Inventory of the California Desert:
A Proposed Methodology". Prepared for USDI. BLM Desert
Planning Program, University of California, Riverside,
CA.

Wilke, Phillip J. (editor)
    1978       Late Prehistoric Human Ecology at Lake Cahuilla,
Coachella Valley, CA. University of California,
Riverside, CA.